MIMESIS
INTERNATIONAL

PHILOSOPHY
n. 71

Elena Ricci

THE VIRTUOUS PATIENT

Cultivating Character Amidst Adversity

MIMESIS
INTERNATIONAL

www.mimesisinternational.com
e-mail: info@mimesisinternational.com

Isbn: 9788869774973
Book series: *Philosophy*, n. 71

CONTENTS

For my mother, father, and Antonino

[E]verything can be taken from a man but one thing: the last of the human freedoms – to choose one's attitude in any given set of circumstances, to choose one's own way. [...] It is this spiritual freedom – which cannot be taken away – that makes life meaningful and purposeful.

Viktor Frankl

INTRODUCTION

Human beings spend a significant part of their lives suffering, being ill, or disabled. Despite this, the possibility of "flourishing" in the context of illness has been largely neglected by philosophers until recent years. This oversight may be attributed to the arguments put forth by several authors, including Aristotle, who posit that flourishing is only achievable when certain preconditions (internal or external goods), such as complete freedom and good health, are met. If we accept this premise as valid, it may appear that flourishing is entirely unavailable to patients, especially those suffering from debilitating chronic or degenerative diseases or illnesses with a poor prognosis. (Bina et al., 2024).

The aim of this book is to challenge this argument. While illness may indeed be a tragic experience, it does not preclude ill individuals from achieving well-being and even flourishing. One can be ill and still benefit from character development, which also plays a role in helping the ill person cope and adapt to new conditions. In summary, I will defend the possibility for patients to experience "illfullness" (i.e. flourishing when one is ill); in this sense, this book aims to provide an "edificationist" reading of illness.

Before delving into the philosophical discussion, a brief outline of the book's structure and the content of each section is presented. This work is divided into three parts. In the first part, the necessity for a "virtue care ethics" framework for the caring contexts will be defended. As will be shown, despite the tendency to regard the ethics of care as the optimal framework for situating care-related discussions (§ 1), this approach exhibits several limitations. Care ethics appears to be a particularly appropriate account due to its focus on topics essential in caring settings, such as the importance of attention, communication, and the universal human need for care. Additionally, it provides a contextual and relational approach, which proves particularly suitable insofar as an ethical framework that is genuinely tailored to the patient's needs requires a certain degree of particularism.

However, these elements, while necessary, are not sufficient. A theory deemed suitable for caring contexts must also adopt a normative outlook, i.e. be capable of guiding the subject into real-life situations, showing what traits they should develop to take care of others. According to Kate Buchanan and colleagues (2022), care ethics aims to be a normative theory, however its normativity is questionable. As will be elucidated later in this work, an approach can be considered normative if it offers guidance for action. This objective is not achieved by care ethics. While it rightly recognizes the centrality of care, it does not provide concrete indications on how to cultivate the right kind of attention, the appropriate degree of proximity, or, more generally, the appropriate way of caring for others.

An additional aspect that makes it unsuitable for use as the sole framework in caring contexts is its exclusive focus on the character of caregivers. In fact, it is crucial for caregivers to possess a "flourished" character to provide appropriate care (Ricci, 2020). However, as I will argue, the character education of the ill person is equally important, as it provides them with the emotional and moral resources necessary to navigate the challenges of illness, fostering resilience, adaptability, and a sense of purpose even in the face of physical limitations or suffering. For these and other reasons, which will be articulated throughout the text, care ethics does not seem capable of functioning as a self-sufficient framework.

Given these limitations, this work will focus on identifying another moral theory that could complement the former by addressing some of its structural deficiencies. The moral theory that appears best suited for this purpose is virtue ethics (§ 2). Virtue ethics, in fact, focuses on character. Thus, it is both concerned with identifying those traits that can be beneficial for various moral agents in different domains of life and committed to character education.

The efficacy of virtue ethics as a robust framework, both from theoretical and normative point of view, has led some scholars to advocate for the reduction of the ethics of care into virtue ethics (e.g. Halwani, 2003; McLaren, 2001; Slote, 1998a;b). However, I will posit that reducing or eliminating care ethics would entail, on the one hand, abandoning the focus on essential topics such as care, attention, communication, and empathy, elements that are at the core of care ethics. On the other hand, maintaining them would require distorting virtue ethics, asking it to become something other than itself. For this reason, I will propose a second path, already partially explored in the literature, namely the possibility of fostering a collaboration between the two moral theories in caring contexts. This would allow for focusing on care without neglecting character development, which is what ultimately enables individuals to adequately care for others. This framework will be referred to as "virtue care ethics". Finally, I will argue that virtue care ethics can serve as a tool not only for healthcare professionals and patients, but also for institutions that aim at being patient-oriented.

In the second part, I will investigate the relationship between suffering, illness, and virtue, showing that despite appearances, these concepts are not mutually exclusive. To do this, I will start by offering some definitions. Initially, the distinction between pain and suffering will be elucidated (§ 3); although these terms are frequently used as synonyms, they refer to different realms. Pain primarily pertains to the biophysical sphere of human beings, whereas suffering is an internal experience and can encompass various dimensions of our lives. Understanding the difference between pain and suffering is crucial, as pain represents only one dimension of need in the life of an ill person. Illness can cause a form of suffering that affects several (sometimes all) dimensions[1], not merely the physical one. While it is undoubtedly true that some traumatic events are so devastating and pervasive that the subject perceives no positive outcome, most forms of suffering offer opportunities for personal development and character growth through trauma processing and positive adaptation.

1 This form of suffering has been defined as "existential suffering" or "total pain".

Indeed, my claim is that illness is not a "suffering trap" as it might initially appear to the patient; it is possible to adapt to it and its consequences through functional coping (§ 4). Thus, if the ill person is supported in the coping process (and in this sense, the support of healthcare professionals, family, friends, and caregivers is crucial), they can adapt and even be positively transformed by it.

Having established the compatibility between the experience of illness and that of flourishing, the third part of this book will explore the significance of a virtuous character as a resource to support patients in more effectively managing their illness and its associated consequences (§ 5). The aim of this section is also to highlight which traits may be best suited to achieve this goal.

Two clusters of virtues will be examined: the "self-regarding virtues" (which include traits such as hope, courage, perseverance, humor, patience, and humility) and the "relational virtues". With respect to relational virtues, I will focus on trust, honesty, and gratitude. As will be discussed in detail, the first set of virtues is primarily oriented toward the ill subject and serves as a set of tools that can help them cope with the various challenges posed by illness. The relational virtues, on the other hand, can facilitate the cared-for[2] to open up and better manage the relationship with both professionals and caregivers/family. This cluster is particularly significant, as caring is inherently relational work, and the subject's receptiveness to caring relationships may result not only in improved levels of well-being but also in enhanced health outcomes.

This part will conclude with some reflections on virtue lists, aiming to highlight both their limitations and their relevance. Indeed, although providing a list of virtues for coping with the experience of illness is a valuable resource, it brings about some problematic issues (§ 6). The limitation of lists lies in the inherent difficulty of establishing a finite set of traits that are universally applicable across diverse social and cultural contexts. Such lists invariably risk failing to satisfy the reader, who may find it necessary to either narrow down or expand it further. However, despite all the problems a list of

2 In care ethics, the term "carer" refers to the individual who provides care, while the care recipient is referred to as the "cared-for" or "care receiver".

traits may have, there is one element which all traits *must* possess to be virtues, and which turns out to be much more important than the traits themselves; this element is virtuosity. As I will show, virtuosity comprises practical wisdom (or *phronesis*) conceptualized as ethical expertise; practical wisdom serves as the *raison d'être* of all virtues, i.e. it is the nuclear element that all "virtues" have in common (De Caro et al., 2025). Indeed, it could be argued that virtues represent "emanations" of practical wisdom, which may manifest as hope in some contexts and courage in others. Thus, the possession of *phronesis* is what enables the subject to understand how they should behave in every situation of life. In this sense, I will embrace a monist version of virtue theory.

One may question the rationale behind enumerating a series of traits if practical wisdom is the sole determinant. In the very last section of this book, I will also highlight the limitations of an approach focused exclusively on practical wisdom. Becoming practically wise requires substantial time and practice, time that not every patient has. Therefore, although this represents the optimal path, and although certain chronic conditions, when addressed early and with adequate character education programs, may allow for this kind of development, it is not always feasible for every patient to pursue such a demanding and long-term process.

In such settings, the cultivation of specific virtues is crucial, as it may enable the ill person to acquire certain traits that can be particularly beneficial, either by counteracting the consequences that a disease may impose on character (e.g. depression or anxiety), or by assisting patients in better managing specific challenges posed by their condition. Furthermore, working on individual traits, beyond a broader approach focused on character development as a whole, may allow patients to concentrate on specific areas of their character.

The challenge for the future remains the implementation of this proposal, namely the development of character education interventions that can serve as a concrete resource for individuals and enable meaningful character work during the course of illness. Although some protocols already exist and the literature attests to their effectiveness, there is still a lack of interventions that move beyond the cultivation of individual traits. These aspects deserve to be investigated and explored further in future research.

This work would not have been possible without the continuous dialogue, guidance, intellectual companionship and friendship of Claudia Navarini. Her thoughtful insights, constant engagement, and unwavering generosity in sharing ideas have deeply shaped the development of this research.

For the stimulating exchanges and thoughtful suggestions that helped me reflect more deeply and revise several sections of this work, I wish to express my heartfelt gratitude to Mario De Caro, Maria Silvia Vaccarezza, Sofia Bonicalzi, and Nancy Snow. Their comments and criticisms have been crucial in refining the arguments and clarifying the overall structure of the text.

I am also sincerely thankful to the members of the Aretai Center on Virtues and to Angelo Campodonico, Luca Iani, Francesco De Vincenzo, Alberto Pirni, Michele Cardinali, Elisa Pelosin, Luca Manfredini, Mario Cagna, Antonio Scoppettuolo, Pierre-Thomas Eckert, Martina Piantoni, Damiano Simoncelli and Wenhui Xie for the enriching dialogue and the exchange of ideas that have significantly contributed to my reflection and to the development of several parts of this book.

Presenting and discussing this work in conferences and workshops has offered valuable occasions for reflection, enabling a more nuanced articulation of several arguments through constructive exchanges.

Finally, I remain profoundly grateful to my family and friends for the patience, love, and care with which they have surrounded me throughout the long process that led to this book.

Claudia Navarini

FOREWORD

It is a privilege to introduce this insightful and timely work, which addresses the often-overlooked potential for human flourishing in the midst of illness. The author, Dr. Ricci, a promising young scholar, engages deeply with key philosophical traditions that examine the relationship between well-being, suffering, and the conditions for flourishing – particularly those developed by Karen Lebacqz, Havi Carel, and Ian Kidd. While some foundational ideas – such as the distinction between pain and suffering – build on a careful engagement with existing scholarship, including the work of Michael Brady and my own, this volume offers a distinctive and original voice, particularly through its development of the innovative concepts of virtue care ethics and illfullness.

This book continues and expands our shared philosophical inquiry into flourishing and virtue ethics with admirable clarity and rigor. At its core, it challenges entrenched Aristotelian (and neo-Aristotelian) assumptions that treat flourishing as dependent on external goods such as health or autonomy. In contrast, the author persuasively argues that flourishing can remain possible even in the presence of illness – a state she terms illfullness. This concept opens an important new space in ethical theory, encouraging us to reconceive well-being in a way that neither ignores the realities of illness nor yields to despair.

The proposal of virtue care ethics is, in my view, a substantial contribution to moral philosophy. By integrating the relational focus of care ethics with the character-centered framework of virtue ethics, the author responds to a pressing need: the development of a normative approach that can guide both caregivers and patients. This combined model deepens our theoretical understanding and has real potential for application in healthcare contexts committed to person-centered care.

Throughout the book, Dr. Ricci thoughtfully clarifies core philosophical concepts. Her exploration of coping and adaptation underscores how virtue development can support positive transformation in the face of illness. This highlights not only the resilience that can emerge from adversity but also affirms the moral agency and dignity of ill individuals.

Particularly compelling is her discussion of two key clusters of virtues: self-regarding virtues – such as hope, courage, patience, humility, and humor – and relational virtues – including trust, honesty, and gratitude. These traits serve as practical tools for patients and caregivers alike, offering ethical resources for navigating the emotional, moral, and social challenges of illness. By articulating how these virtues can manifest within the experience of chronic sickness and vulnerability, the author advances a nuanced and deeply human account of moral development in caregiving relationships.

The book also provides an important reflection on the role of practical wisdom (*phronēsis*) as the unifying ethical expertise. Drawing on the Aretai Model of practical wisdom, the author emphasizes that while achieving *phronēsis* is a lifelong process – perhaps unattainable for many patients – its central role in virtue theory remains crucial. She carefully balances this ideal with a practical view: in contexts where time and capacity are limited, cultivating specific virtues may provide a more attainable path to ethical growth.

Finally, Dr. Ricci's interdisciplinary and compassionate approach addresses pressing challenges faced by both patients and healthcare systems. Her call for a broader understanding of palliative care – one that extends beyond end-of-life contexts to encompass chronic illness and disability – underscores the practical and moral relevance of her work. These insights contribute meaningfully to ongoing debates about care, suffering, and well-being, offering a hopeful path forward grounded in ethical development and human dignity.

Overall, I believe this book represents a significant contribution to contemporary moral philosophy and medical ethics. I am confident it will become a vital resource for scholars, healthcare professionals, and institutions seeking to better understand and support the possibility of flourishing amid illness.

PART ONE
THE NEED FOR A "VIRTUE CARE ETHICS" IN CARE SETTINGS

The importance of care in ethical discussions has become increasingly evident over time, especially in care settings. Traditional moral theories which evaluate good actions based on universal principles (like deontologism) or the calculation of outcomes (like utilitarianism), appear insufficient to capture the nuanced, relational dimensions of such contexts. Care ethics, by contrast, shifts the focus from abstract norms to the interpersonal connections, emphasizing the moral significance of issues that have been overshadowed by other moral theories such as the importance of the quality of healthcare professional-patient relationship as well as dispositions such as respect, trust, and empathy (e.g. De Panfilis et al., 2019). Understanding the importance of these factors is crucial for practitioners to incorporate them into their practice. Indeed, we cannot expect a doctor, for example, to allocate time to communication or endeavor to become an effective communicator without acknowledging and understanding its importance.

While care ethics has been valuable in reframing ethical debates, it has limitations. Indeed, in care contexts it is vital not only to understand the significance of effective communication and to identify the professional's character traits that would most benefit patients, but also to support physicians in enhancing their communication skills and to assist them in developing positive traits. However, as will be discussed, care ethics lacks a robust normative foundation, which undermines its capacity to provide clear guidance for action. It is precisely for this reason that in this first part of the book, I will argue in favor of an integrated approach, highlighting the valuable collaboration between care ethics and virtue ethics. Virtue ethics focuses on the concept of character and the achievement of a good life across all domains of existence, thus emphasizing character education. Through the integration of the relational emphasis inherent in care ethics with the character-driven approach of virtue ethics, it becomes feasible to establish a more comprehensive framework that not only underscores the moral significance of care but also equips healthcare professionals with the virtues necessary to deliver effective, compassionate, and ethically grounded care.

CHAPTER 1
CARE ETHICS
History, Definitions and Problems

1. *Historical background*

To comprehend the importance of the virtue care ethical framework in health-care contexts, it is necessary to first delve into the origins and fundamental characteristics of the ethics of care. Elucidating its historical foundations, theoretical premises, and distinctive features will facilitate a more profound assessment of the advantages and limitations of this moral perspective, thereby illuminating its considerable relevance in contemporary ethical discourse.

Care ethics (or ethics of care) is a moral theory that has undergone various formulations over time. The earliest traces of this ethical paradigm can be found in the writings of some important thinkers of the 1700s and 1800s such as Mary Wollstonecraft, Catherine and Harriet Beecher, Charlotte Perkins, Elizabeth Cady Stanton, and Lucretia Mott. However, the emergence of care ethics as an independent moral theory is commonly referred to works originating in 20th-century feminist thought[1], particularly those of Carol Gilligan (1982) and Nel Noddings (1982; 2002; 2013)[2].

---

1 It should be made clear that although the birth of this philosophical current is deeply linked to 20th century feminist thought, it is not to be understood as synonymous with feminist ethics, nor should it be reduced to the latter.

2 Authors who have most influenced the emergence of the ethics of care also include Virginia Held, Eva Feder Kittay, Sara Ruddick, Joan Tronto. I recall below some of the philosophers' most important works, Held, 1993; 2006; Kittay, 1999; Kittay & Mayers, 1987; Ruddick, 1982; 1989; Tronto, 1993. With regard to other authors who have fostered its development, it is essential to mention Michael Slote (see 1988a; 1988b; 2007) and Stan van Hooft (especially, 1995; 1999).

In addition to feminist contributions, Milton Mayeroff's short book *On Caring* has significantly shaped its theoretical articulation (Mayeroff, 1977). Although Mayeroff does not elevate the philosophy of care to the status of a moral theory, *On Caring* contains all the elements that subsequent theorists have deemed essential, such as the centrality of caring or the theme of mutual dependency among human beings.

From a theoretical standpoint, care ethics posits its foundational principles on the primacy of care, which is regarded as the heart of human existence and the fundamental element for the development and flourishing of every individual. Indeed, according to theorists, since human existence is marked by an ontological limit (Heidegger, 1996) – and it is thus vulnerable and fragile – human beings require gestures of care throughout the entire course of their life. This constant need to be cared-for and the memory of having been cared for by someone engenders a natural inclination to provide care for others. This natural tendency can also be associated with the moral responsibility that individuals feel towards others. According to care theorists, because we are fragile, vulnerable, and dependent, there exists a commitment to care for one another. Luigina Mortari and Marco Ubbiali writes about this:

> We need others as much as others need us. The answer to the other's "neediness" is care. The fact that everyone needs care makes self-evident that we must take care of the others. It is the continuous exchange of care that makes life possible. Coexisting implies to take care of the others. In this sense, taking care of the others is not just a possible ontological perspective, but a necessity for the whole existence (Mortari & Ubbiali, 2017, p. 270).

Central to the ethics of care is thus the concept of "interdependence", or mutual dependence. In this regard, care ethics diverges from deontological and consequentialist ethics due to its relational orientation (Noddings, 2013) and a focus on the subject's emotional rationality (see Scoppettuolo, 2017). Care theorists posit that the morality of an action is not determined by rules, maxims or principles. It rather is rooted in their capacity for care. In other words, the ethics of care suggest that a caring action cannot be predetermined as good based on specific rules, maxims, and norms. Conversely, since every situation is unique, it requires a situated choice; it is the quality of care provided that determines the moral value of the action. The morality of an action, therefore, is not defined by the application of a universal rule, but by the quality of the care provided to that person within that specific context.

The characteristics of this model, which is both contextual and relational, render care ethics applicable across a variety of care settings, including educational, political, social, and health contexts. With regard to the medical-health field, care ethics has become increasingly associated with the philosophy of medicine, bioethics, nursing ethics and palliative care ethics (see van Hooft, 2003). This association is not unexpected. The latter domain clearly illustrates the applicability of care ethics in healthcare. In palliative care, attention to relational aspects and communication, as well as the recognition of patients' emotions, desires, and needs, are essential for providing adequate care[3].

2. *Nature and Purposes of Care*

Having elucidated the substantive characteristics of care ethics, and its significance in relation to medical ethics, it is necessary to explore how care is conceptualized in the literature. In this regard, Daniel Engster offers a clear articulation of how care is commonly understood, identifying it's three "basic aims":

> First and foremost, we care for others when we help them to satisfy the basic biological needs necessary for survival and basic functioning. Basic biological needs include needs for food, sanitary water, clothing, shelter, rest, a clean environment, basic medical care and protection from harm, as well as the need at least among infants and children for physical contact and holding. [...] The second aim of caring is helping

3 A key figure who demonstrates the connection between care ethics and medical ethics is Joan Tronto. Tronto identifies four "elements" which are essential to caring relationships: attention, responsibility, competence, and responsivity. These components are widely referenced in the ethics of palliative care. See Navarini, 2020, p. 75.

> others to develop or sustain their basic capabilities for sensation, emotion, movement, speech, reason, imagination, affiliation, and in most societies today, literacy and numeracy. [...] The third basic aim of caring is helping individuals to avoid or relieve suffering and pain so that they can carry on with their lives as well as possible (Engster, 2005, pp. 50-53).

Care, in this sense, can be understood as performing three core functions: it responds to individuals' fundamental physical needs, supports the development and preservation of their basic human capacities, and seeks to reduce or prevent suffering in order to promote a dignified and livable existence.

While in the literature there is generally a consensus regarding the purposes of caring, the same cannot be said about its nature. There is in fact a cleavage between those who define care as a practical activity or work (e.g. Held, 2003; 2006; Kittay, 1999; 2002; Mortari, 2016; Mortari & Ubbiali, 2017; Ruddick, 1982; 1989) – specifically, a practical activity that invariably requires commitment and some *form of relationship* – and those who define it as a *core virtue* (Curzer, 1993; Engster, 2005; Halwani, 2003; McLaren, 2001; Navarini & Pijas, 2025; Rachels, 1999; van Hooft, 1995). Care is defined as a virtue[4] when the focus is placed on the internal dispositions of the carer, i.e. on their character. In this sense, it is thus understood as an acquired and stable trait that informs moral conduct in a sustained and coherent manner. Conversely, philosophers who define care as a practical activity tend to describe it as a particular form of "relational labor" (or a cluster of practices) (Held, 2006). This approach to care emphasizes that both character and commitment on the part of the caregiving person are necessary but not sufficient; for a caring action to be effective, a response from the care recipient is also necessary. Simple gestures, like a smile, can indicate whether the carer's actions are effective, allowing them to adjust their approach based on the needs of the cared-for. Held writes about this:

> Care must concern itself with the effectiveness of its efforts to meet needs, but also with the motives with which care is provided. It seeks

4 For a review, see Steyle, 2019.

> good caring relations. In normal cases, recipients of care sustain caring relations through their responsiveness – the look of satisfaction in the child, the smile of the patient. Where such responsiveness is not possible – with a severely mentally ill person, for instance – sustaining the relation may depend entirely on the caregiver, but it is still appropriate to think in terms of caring relations: The caregiver may be trying to form a relation or must imagine a relation (Held, 2006, pp. 36-37).

Regarding the two hypotheses – whether care is a relational labor or a virtue – it resembles more a (relational) activity than a single disposition. Commitment and a flourished character are insufficient for an action to qualify as caring; as I will argue, its success requires the active involvement of both the carer and the cared-for[5]. However, since commitment to practice is character-dependent, it is essential that both parties possess an educated character. Consider, for instance, the relationship between a nurse and their patient: the nurse may possess extensive knowledge and experience, along with a compassionate and caring disposition to take care of their patients. However, the efficacy of the nurse's care is significantly dependent on the patient's cooperation and engagement in the therapeutic process. If the patient is unwilling to communicate their needs, adhere to prescribed treatments, or remain receptive to the nurse's efforts, how can they take proper care of that patient?

In this sense, care could resemble a "relational virtue"[6]. Relational virtues can be manifested when there is a relationship between two individuals; in this regard, an example is *filial piety*. However, the defining characteristic of any trait, including a relational trait, is that it delineates a particular course of action. This principle, however, does not apply to care. When philosophers refer to the "virtue of caring", they typically allude to a set of dispositions that facilitate the subject's ability to understand the person's needs, to take charge of them, and respond accordingly. Nevertheless, these

5 In this sense, it is also crucial to encourage the cared-for to work on their inner world.

6 Relational virtues are defined as a "subcategory" of virtues which are generally considered as "required for an individual as an excellent participant of the given type of intimate relationship" (Um, 2021, p. 96).

processes require the coordinated action of several virtuous traits. As Navarini and Pijas write: "care must be properly *grounded* in someone's character" (Navarini & Pijas, 2025, p. 10). Care, therefore, encompasses a broader scope than other virtues. Consider, for instance, the courageous firefighter, who is characterized by a willingness to enter a burning building to rescue trapped individuals. They will evidently require several other virtues to perform that role effectively. However, it is relatively straightforward when discerning the type of action required when confronting fire to isolate courage. Consider now a physician who is tasked with communicating an unfavorable diagnosis. In this scenario, to take good care of a patient requires multiple traits and not just one. Care, in fact, can be expressed in many ways depending on the context and it therefore necessitates the collaboration between different virtues. It is precisely the cultivation of those traits that enables the carer to care for the other.

The idea that care, rather than a virtue, is a relational labor for which a number of traits are needed is widely supported in the literature. For example, Mayeroff, in his *On Caring,* dwells on the motivational aspects that characterize caring, which he does not define as a unitary virtue, but rather as a series of traits – namely, patience, honesty, trust, humility, hope, and courage – that would represent "the major ingredients of caring" (Mayeroff, 1971, p. 9). Similarly, Engster (2007) has also pointed out that a practice cannot be considered as caring *per se*; rather, it can become so according to its aims and its virtues. The philosopher identifies three core virtues of care. He identifies attentiveness – or what Lawrence Blum refers to in his work as "moral perception" (Blum, 1994) – which corresponds to "perceiving" the need of the other person and responding appropriately to it[7]. According to Engster's view, without *attentiveness* the actions of the carer would remain mere intentions, lacking both an understanding of the other's needs and the ability to respond appropriately. The second virtue identified by Blum is

7 Noddings defines receptivity with the term "engrossment". Engrossment is, according to the philosopher, "an open, nonselective receptivity to the cared-for" (Noddings, 2005).

responsiveness, which refers to the ability to engage in dialogue with the other and to monitor their reactions to the care provided. The last virtue is *respect,* which is the recognition of others as morally worthy of both attentiveness and responsiveness.

Mortari and Ubbiali also argue that care is a practice which "reveals itself through ways of being" (Mortari & Ubbiali, 2017, p. 270), i.e. by paying attention to others, listening, and being present. The core of such ways of being is, according to the authors, a set of traits such as courage, respect, and generosity[8].

Thinking that care is not a unitary trait but that it consists in a series of dispositions could lead us to think of it as a meta-virtue, that is, a virtue which we can define as an "enabling disposition". However, it should be emphasized that care is not a disposition that "nurtures" all the other virtues by making them operative. On the contrary, care depends on the exercise of the other virtues. It is a cluster of traits – always different according to context – that, if properly cultivated, can make a certain practice a care practice.

Caring must be thus embedded in a virtuous character to be considered virtuous as it entails caring about the right things, in the right way, and with the right dispositions. Peter Allmark offers a particularly insightful perspective on the relationship between caring and virtue: "Caring is not a virtue, but having a virtue involves caring about the right things in the right way" (Allmark, 1998, p. 467). The engine of action remains thus the agent's character, since it is precisely our internal dispositions that help us in understanding how to act: a *virtuous* teacher is committed to caring for their students, but they can have *such care* precisely because they are a *virtuous* teacher. The same thing applies to the virtuous doctor, the virtuous nurse, the virtuous parent, the virtuous patient, and so on.

Ultimately, we can conclude by saying that caring represents a relational practical activity that requires the carer to develop several traits to be a virtuous care. However, since caring is inherently relational, its effectiveness depends on the active collaboration

8 As will be noted, the virtues mentioned by Mortari and Ubbiali recall quite clearly those proposed by Mayeroff.

between the carer and the cared-for. As we will explore later, the moral development of the cared-for also plays a fundamental role in this dynamic.

3. *A Few Problems*

As previously discussed, over time, the ethics of care has become closely aligned with the medical field, and it is increasingly regarded as a key resource for supporting clinical practice[9].

Vicki D. Lachman for example underscores the importance of applying care ethics in nursing practice (Lachman, 2012). In *Applying the Ethics of Care to Your Nursing Practice*, she underlines the centrality of the nurse-patient relationship, asserting that "good care" constitutes a moral responsibility within the nursing profession. According to Lachman, effectively translating this ethical orientation into clinical practice requires nurses to engage in continuous education and professional development[10]. Specifically, they should focus on acquiring skills that would allow them to recognize and respond appropriately to the physical, psychological, spiritual, and social needs of their patients[11]. Lachman thus advocates for the implementation of the ethics of care in clinical practice, highlighting its fundamental role in shaping meaningful nurse-patient relationships and in orienting ethical decision-making within the caregiving process.

Another significant contribution supporting the relevance of care ethics in healthcare is offered by William T. Branch, who develops what he terms "the ethics of patient care", stressing the central role of care ethics in medical practice. Branch specifically proposes

9 Authors who consider care ethics as a key resource for caring practices include for example Branch, 2015; De Panfilis et al., 2019; De Vries & Leget, 2012; Hermsen & Ten Have, 2003; Inge van Nistelrooij et al., 2017; Lachman, 2012; Schuchter and Heller, 2018.

10 Lachman also stresses the importance of practitioner's self-care in caring practices. According to her, this trait is important to have the right energy and motivation for taking good care of patients.

11 In particular, Lachman refers to Tronto's theoretical proposal.

an approach grounded in the physician-patient relationship and in traits such as compassion, respect, humility, honesty, and integrity (Branch, 2015)[12]. According to Branch, this model is essential for physicians to fulfill their primary duty: to provide the best possible care.

To facilitate the adoption of this framework, he emphasizes the importance of cultivating moral development throughout medical education, highlighting the intrinsic connection between ethical conduct and high-quality clinical practice. It is, in fact, ethical behavior that enables physicians to act competently as according to Branch, ethical behavior comprises "moral sensitivity (the ability to recognize ethical issues), moral commitment (the determination to do what is right), and moral behavior (skills at implementation), with moral reasoning (being able to weigh the rights of others and the principles at stake)" (Branch, 2000, p. 505). All these aspects are core to medical practice.

Ludovica De Panfilis and colleagues also defend the relevance of applying the ethics of care to clinical settings. To explore this, they conducted a qualitative study aimed at understanding whether and how the ethics of care informs the way professionals perceive and manage ethical issues in palliative care (De Panfilis et al., 2019, p. 1). The study's findings reveal that, although the interviewed healthcare professionals did not explicitly refer to a particular ethical framework, they tended to approach ethical aspects of their clinical practice and moral dilemmas using notions and concepts from the toolkit of the ethics of care[13].

De Panfilis and colleagues also noted that healthcare professionals struggle to address ethical issues without becoming overly involved, particularly on an emotional level, and to navigate moral dilemmas effectively. To mitigate this, they advocate for a "step by step moral

12 In this sense, the philosopher already shows us the fundamental importance of an integrated approach between the ethics of care and virtue ethics.

13 De Panfilis and colleagues write about this: "HPs [health-care professionals] talk about ethical issues in palliative care using notions and concepts such as caring relationship, listening, dialogue. These aspects are strongly highlighted in the ethics of care approach" (De Panfilis et al., 2019, p. 7).

training" approach designed to equip healthcare professionals with practical tools for managing both minor and significant ethical challenges encountered in daily practice. The authors suggest that such moral training should be also grounded in the ethics of care, which they consider to be fundamental for practitioners. According to De Panfilis and colleagues, care ethics would help professionals in recognizing the significance of their relationships with patients. It would also help them to attribute the right meaning to such elements as listening, dialogue, telling the truth, empathy, and respect for the patient's autonomy (see De Panfilis et al., 2019, pp. 7-8).

The authors also underscore the importance of integrating the ethics of care with other moral theories. Specifically, they highlight principialism and virtue ethics as complementary frameworks. Principialism would guide healthcare professionals by offering key principles (autonomy, beneficence, non-maleficence, and justice) (Beauchamp & Childress, 1994) that assist in reasoning and responding to moral dilemmas. Virtue ethics, on the other hand, could help them develop appropriate "attitudes" and traits, especially practical wisdom, which is essential for navigating complex ethical situations. Indeed, De Panfilis et al. conclude their work arguing in favor of an ethical training that integrates the three moral theories[14], consistently underscoring the foundational role of the ethics of care.

Lachman's, Branch's, and De Panfilis and colleagues' works bring to light common elements and offer key insights about why the ethics of care is such a useful tool in clinical settings. Care ethics, however, is not a resource without pitfalls that lie behind what I may

14 Talking about moral training, De Panfilis and colleagues believe that the latter is crucial in that it would be able to help healthcare professionals in developing the appropriate skills to recognize moral dilemmas, to deal with major ethical issues, and to understand and address patients' needs. This would also enable them to make decisions without the emotional involvement that can sometimes be a major limitation, especially when a particularly complex choice needs to be made (De Panfilis et al., 2019, p. 8). The authors do not explain how this ethical training should be structured, nor do they demonstrate its effectiveness. However, the possible integration of these three approaches into a practical framework remains an interesting suggestion and would deserve to be developed into a concrete proposal.

call some “structural deficiencies”. In the next section I will analyze the four problems that undermine the self-sufficiency of this theory, namely the *systematicity problem,* the *normativity problem,* and the *unidirectionality problem.*

3.1. *The Systematicity Problem*

A first problematic aspect that can be found in the ethics of care is related to the little (if any) focus on character education. As previously discussed, caring is inherently a relational practice that necessitates collaborative involvement of both the carer and the cared-for. However, for healthcare professionals to provide adequate care (and thus understanding patients’ needs, responding with genuinely caring actions without paternalistically imposing their own notions of good, and being motivated by appropriate motivation, etc.), it is necessary to work on their own character.

Care ethicists have been concerned with identifying the carer’s essential dispositions and have acknowledged the importance of practice in character development. Noddings for example writes in this respect that “there is a form of competence in caring [...] One must have encounters, legitimate opportunities to care, in order to go about caring effectively” (Noddings, 2013, p. 123). Similarly, Stan van Hooft emphasizes the subject’s moral growth through practice (van Hooft, 1995). However, a precise account of what moral development entails remains largely absent from care ethics literature.

This lack may be attributed to the non-systematic nature of care ethics. Many care theorists maintain that we do not require structured guidance to appropriately care for others, as this ability arises naturally from the experience of being cared for. Yet even assuming that the experience of being cared for nurtures a sense of responsibility and a willingness to care for others, this does not necessarily mean that one is also capable of delivering good care. A novice drawer, for example, will never become an expert unless they engage in regular practice, have models to follow, submit their sketches to the judgment of experts, and critically revise their work. Similarly, if Mark desires to be a kind person, he will not develop the virtue of kindness if he is unwilling to accept any criticism and if, for instance, he believes he already knows how to be kind. Practice must be directed toward improvement; without guidance and reflection, the carer risks acting on the basis of sentimentalisms[15] and intuitions. However, feelings do not constitute a reliable guide in moral action, while intuition, as will be discussed in the following chapters, becomes a trustworthy form of moral discernment only once a high level of moral expertise has already been achieved (Stichter, 2018a).

3.2. *The Normativity Problem*

The systematicity problem gives rise to a further challenge: the lack of normativity. A non-systematic theory, by definition, cannot function normatively as it lacks the structural coherence necessary to offer consistent guidance for moral action. Although the ethics of care aspires to orient caring practices, its pronounced "particularism" renders it an ambiguous framework (e.g. McLaren, 2001; Rachels, 1999).

If adopted as a primary ethical reference, the ethics of care would require practitioners to determine the most appropriate course of action in each specific case, drawing upon the particular characteristics of the cared-for, the context, and a selection of virtues deemed relevant in that moment, without having been properly trained or educated to do so. As previously discussed, the ethics of care does not provide professionals with clear guidance for moral development. This absence of structured ethical framework for character work can leave caregivers reliant on personal intuition and emotions, which may not always suffice for ethical decision-making, especially when faced with complex dilemmas.

Empirical studies, such as the one conducted by De Panfilis et al., demonstrate that healthcare professionals often struggle when confronted with moral dilemmas, and are often unable to ensure

15 In this sense, the ethics of care certainly draws many elements from other perspectives such as that of David Hume (see Hume, 1978). In fact, as Slote states, "the ethics of care is historically rooted in the moral sentimentalism of Shaftesbury, Hutcheson, Hume, and Adam Smith" (Slote, 2007, p. 3). For this reason, it can be considered a form of sentimentalism (Slote, 2007, p. 105).

proper proximity to their patients, a proximity that avoids both over-involvement[16] as well as shortcomings of care that are typical of "false caring". In order to address this type of ethical vulnerability, alternative frameworks such as principialism are structured around sets of defined principles or rules. Such frameworks recognize the need for moral agents to be guided in their reasoning and choices. Virtue ethics, on the other hand, while it similarly rejects the prescriptive rigidity of principles and aligns more closely with the particularism of care ethics, does not merely identify the virtues required for a specific moral domain, it also provides an account of how those virtues are to be cultivated. Virtue ethics thus aims to form autonomous moral agents. By contrast, the ethics of care, by rejecting both rule-based principles and the structured cultivation of character, risks being unable to offer meaningful guidance for action. In doing so, it may leave professionals vulnerable to the instability of their own emotions, ultimately undermining the moral efficacy of their actions, even when motivated by the best of intentions.

3.3. *The Unidirectionality Problem*

A further and significant limitation of the ethics of care has been insightfully identified by Navarini (2020). Navarini notes that even when care ethics attempts to identify those character traits that contribute to making a practice ethically sound, it does so in a unidirectional way, focusing almost exclusively on the carer. The cared-for, by contrast, is typically portrayed merely as a passive recipient of care, rather than as a fully-fledged moral agent. In this sense, the ethics of care tends to treat patients as if nothing should be expected of them (Kidd, 2012). Their condition of vulnerability is implicitly taken as a justification for exempting them from any moral growth (see Navarini, 2020, p. 94). Indeed, the only role that care ethicists seem to assign to the cared-for is to facilitate the work

16 Excesses of care can expose caregivers to the risk of developing conditions such as burnout syndrome. On burnout syndrome in palliative care see e.g. Martins Pereira et al., 2011.

of health-care professionals, and to contribute to the care process by offering feedback.

This unidirectional view of the caring relationship, which is also present in other models such as the biomedical approach of Pellegrino and Thomasma (1993), constitutes a serious limitation for the ill person who should be supported in a path of flourishing. Supporting patients in their moral growth is crucial for several reasons. First, insofar as care unfolds within a relational context, the cultivation of character may enable patients to approach the care relationship with greater openness and receptivity. A further reason is that a virtuous character represents a key resource for coping more effectively with the experience of illness, as will be explored further in the third part of this work. As Lebacqz writes, illness often enters life as an "unwelcome intruder", disrupting the subject's narrative and bringing about profound changes (Lebacqz, 1985). Adapting to this new condition is possible, but such adaptation cannot be taken for granted. Patients often respond to the communication of the diagnosis by reporting negative emotions or even depressive symptoms, which can impede their moral growth, diminish their sense of hope, and foster feelings of worthlessness (e.g. Chochinov, 2001). For this reason, recent literature underscores the importance of character work when facing illness, highlighting how the cultivation of virtues can help patients to navigate the challenges of their condition more effectively. As Ian Kidd writes:

> It is not difficult to see how virtues enable an ill person to better cope with their illness. Consider a series of typical virtues. An adaptable and creative person can better respond to their illness than an inflexible, increative person, especially if they are also patient and have fortitude. Confirming to drug regimens and exercise requires discipline, while engaging with healthcare professionals invites trust and honesty, as well as hope concerning possibilities for future health and gratitude for what health one does have (Kidd, 2012, p. 502).

A strengthened character can indeed enable the ill person to re-signify their experience and to better cope with their new condition and with the small and large daily challenges that it poses. This highlights the need for an ethical framework that integrates attention to the patient's moral character and actively supports their development along a path of personal growth during the illness trajectory. We will return to this point in the next part of this work.

CHAPTER 2
BENEFITS OF AN INTEGRATED APPROACH

In the preceding paragraphs, we have seen why the ethics of care is often regarded as a highly appropriate theoretical framework for caring contexts. The ethics of care reminds us that medicine is not solely a matter of clinical decisions. It also fundamentally concerns the quality of the caring relationships which consist of communication, responsivity, empathy, moral concern, and sensitivity. However, as I have argued, these elements risk remaining ineffective in practice if the carer does not engage in the cultivation of their own character. Indeed, it is not sufficient to know that caring requires attentiveness in order to become attentive; nor is the sporadic practice of certain positive traits enough to ensure that one is attentive to the right things, in the right way, and for the right reasons.

Additionally, for an action to be caring, it must be directed toward a person who is open to be cared for, and willing to cooperate in the caring relationship. It is difficult, for example, for a physiotherapist to take care effectively of a patient who frequently misses appointments and fails to engage daily with the proposed exercises – either because the exercises do not produce immediate results or because the patient has lost faith in the possibility of improvement. In such cases, the patient would benefit from cultivating certain character traits, such as perseverance, courage, and hope who would help them adhere consistently to the therapeutic plan, and trust, which would foster committed engagement with the physiotherapist's guidance. From this perspective, it becomes evident that while the ethics of care is a necessary moral framework, it is not self-sufficient[1], especially

1 Indeed, in this regard Gilligan suggested the need to combine care ethics with other ethical frameworks (Gilligan, 1982).

when the goal is to translate ethical theory into effective practice. The centrality of care, along with a relational approach grounded in the analysis of the person's needs and the specific context in which the person is situated is clearly evident. Yet, it remains crucial to provide practitioners with a normative and systematic framework that supports the cultivation of character traits conducive to a form of caregiving that is not only technically effective, but also ethically grounded. Good care practices, in fact, depend not solely on clinical or technical competence, but also on the caregiver's caring manner and thus on their character (Ricci, 2020).

It is important to emphasize here that this process involves self-examination and the recognition of one's own limits and needs. To take care of another person presupposes the capacity to recognize and engage with one's own vulnerabilities and limits; caring for others is always, to some extent, a form of caring for oneself. Accordingly, the virtuous practitioner will be able to discern the appropriate degree of proximity in the caring relationship, while remaining attuned to their own limits and needs.

On the other hand, since care is a relational activity, patients should be supported in moral flourishing. Working on one's character can be a deeply beneficial part of this process, as it may help individuals to develop virtues and character strengths that may enhance their ability to engage in caring relationships and to cope more effectively with the challenges posed by illness.

However, as we have seen, care ethics lacks a focus on character, and in particular on character education. Precisely for this reason, what I propose is an integrated approach that brings together the ethics of care and virtue ethics into a new ethical framework, which I will refer to as "virtue care ethics". This integrated model aims to offer a systematic and normatively grounded theory that simultaneously prioritizes the moral development of both the carer and the cared-for.

1. *Integrating Care Ethics and Virtue Ethics in Caring Practice*

The idea that a collaboration between care ethics and virtue ethics would be fruitful in applied contexts is present in the literature. For example, Mortari and Ubbiali have strongly endorsed the advantages of an integrated approach between these two moral theories. In 2014, they launched an ethical project named *MelArete*, a term formed by combining *mele* – an ancient Greek word associated with care – and *arete*, the classical term for virtue. *MelArete* is a research initiative aimed at educating children in virtue, with the ethics of care serving as the foundational framework (Mortari & Ubbiali, 2017).

Mortari and Ubbiali emphasize that any robust moral education should recognize the primacy of care. As they write: "the pedagogical theory of MelArete states that in order to develop a project that is in relationship with the core of life we must educate to care" (Mortari & Ubbiali, 2017). In their view, care constitutes the core of human existence and is the principal condition for human flourishing. Since human beings are inherently vulnerable and fragile, they cannot flourish without care.

Mortari and Ubbiali conceptualize care as a practice. However, they hold, to effectively educate individuals about care, it is essential to comprehend its defining characteristics. As a practice, care manifests through what they term "ways of being" (such as paying attention, listening, being present, and showing empathy). These ways of being are underpinned by virtues like respect, generosity, and courage. Consequently, care ethics and virtue ethics appears to be deeply interconnected:

> Care, since it is a practice, reveals itself through ways of being: to pay attention, to listen to, to be present, to act with delicacy, to pronounce words that attest a thoughtful consideration for the other, to show empathy, etc. But at the core of these ways of being it emerged to be intimate ethical postures as to have respect, to be generous, to have courage, to feel oneself responsible for the other: since these postures can be identified as virtues, it is legitimate to state that the intimate core of care is an ethics of virtues. From this assumption it emerges that educating to care means educating to virtues (Mortari & Ubbiali, 2017, p. 276).

While Mortari and Ubbiali do not explicitly label their pedagogical approach as virtue care ethics their work within the MelArete project embodies this integration.

Another contribution that underscores the value of integrating care ethics and virtue ethics – particularly in the moral education of healthcare professionals – is the work of De Panfilis and colleagues (2019). As previously noted, the authors argue that healthcare professionals require a structured step-by-step moral training. Such training is crucial not only to help practitioners develop the capacity to address complex ethical dilemmas but also to prevent potentially dangerous attitudes, such as excessive emotional involvement, which are common in ethically demanding situations. This training should encompass care ethics, virtue ethics, and principialism: care ethics emphasizes the recognition of the centrality of relational engagement and empathy; virtue ethics, on the other hand, assists practitioners in developing the appropriate character traits and cultivating practical wisdom (*phronesis*); and principialism offers guiding principles to be applied when facing moral dilemmas.

While this book primarily advocates for the relevance of a virtue care ethics approach in clinical contexts, it is nevertheless important to acknowledge that De Panfilis and colleague's step-by-step moral training framework may prove valuable in certain contexts. Although virtue care ethics may represent a key resource in care settings, the development of moral character and the ability to recognize the ethical dimensions of a situation take time. It requires learning how to regulate one's emotions[2] and cultivating the kind of moral sensitivity that allows to discern the most appropriate response in a given context. For this reason, especially at an early stage in their ethical development, practitioners may benefit from having at least a set of reference values or principles. Being introduced, through formal training, to key moral principles and to the evaluative processes necessary to determine which of them should be prioritized can be a meaningful first step. However, such an approach cannot be considered sufficient in the long term, given that principles often come into conflict with one another. In dilemmatic situations, for example, it may be hard to determine the boundary between respecting autonomy and ensuring justice. Similarly, it may be difficult to understand when a commitment to

2 On the ability of virtues to regulate emotions see e.g. Darnell et al., 2019; Snow et al., 2019.

honesty might actually harm the patient. Nonetheless, considering that healthcare professionals may lack adequate bioethical training, providing them with an initial framework based on the centrality of care and on the ethical principles can offer useful ethical scaffolding. Having laid this groundwork, training can gradually integrate a character-based approach.

De Panfilis and colleagues do not provide clear operational guidance on how to construct such an integrated moral training pathway. This remains an open area for future research, which should aim to develop structured ethical protocols specifically designed for clinical practice.

2. *The Possible Contribution of Virtue Ethics*

Building upon the insights from the MelArete project and the step-by-step ethical training proposed by De Panfilis et al., it is evident that while care ethics provides a foundational framework in care contexts, it may not be entirely self-sufficient. For this reason, it needs to be integrated with other moral theories. I suggest that virtue ethics represent a valuable resource in this sense.

Unlike deontological approaches that focus on adherence to norms and maxims, virtue ethics posits that moral action depends on a flourished character (Campodonico et al., 2017, p. 10). A virtuous character supports the agent in deliberating and acting well across diverse moral domains and contexts, shaping their identity, motivation, moral sensitivity, and emotions. This approach is particularly valuable in care settings; cultivating virtues can help professionals manage emotions, make appropriate decisions in complex situations, and discern the appropriate level of support to provide to patients, thereby enhancing the quality of care provided. Virtue ethics may facilitate this through a systematic approach that not only identifies the most relevant character traits in a given context but also provides practical methods for cultivating them[3].

---

3 Although few character education protocols currently exist for healthcare professionals, some preliminary initiatives have been proposed in the literature. With targeted research, these could be further investigated to

Up to this point, I have discussed the advantages of integrating the ethics of care with virtue ethics from the caregiver's perspective. However, as previously noted, caring is inherently relational and thus depends not only on the caregiver's actions but also on the attitudes and dispositions of the cared-for, that is, on their character. While the ethics of care has traditionally focused on caregiving figures, the moral character of the ill person has become the center of increased attention in virtue ethics (e.g. Brady, 2018; Carel & Kidd, 2020; Navarini, 2020). Although this ethical framework has long centered on healthcare professionals' dispositions, there is a growing recognition of the relevance of a virtuous character in the experience of illness. In particular, the literature has shown that character strengths and virtues may support functional coping and facilitate positive adaptation. This may occur, as psychological studies suggest, because virtuous individuals appear to be more inclined to identify positive elements within their experiences, even in adverse and traumatic situations (e.g. Ruini & Vescovelli, 2013).

To better grasp the significance of virtues and moral flourishing for the ill person, consider the following scenario. Let us imagine two patients diagnosed with the same rare form of cancer. The first, Peter, is a deeply distrustful person who has rarely taken risks in his life and often struggles with a tendency toward sloth. He expresses little hope for remission and firmly believes that he will ultimately succumb to the disease. The second patient, Mary, is a former prima ballerina with the American Ballet Theatre. Her professional experience has taught her that achieving meaningful goals requires discipline, perseverance, and a willingness to endure hardship. She applies this approach to every challenge in her life, including that of illness. Mary is also characterized by a great vital instinct, she places deep faith in both God and medical science, and is confident that her physician will find a cure for her condition.

Let us now assume that an experimental therapy has recently been developed, and both patients are offered the opportunity to undergo the treatment. This new protocol presents a modest chance of success

assess their potential and limitations, and subsequently refined or expanded (See, for example Seoane et al., 2016).

(approximately 30%) along with the risk of side effects. Despite the low probability of remission, the oncologist proposes the therapy to both individuals as they are young, and he believes their bodies could withstand the treatment and that they may derive significant benefit from it.

If you were told that one of them accepts the treatment and subsequently experiences full remission, whom are we more inclined to identify as the likely candidate, Peter or Mary?

Let us then consider a second example involving two patients, Hannah and Mark, both diagnosed with Parkinson's disease at the same clinical stage. They present with comparable symptoms, share similar socioeconomic conditions, and both benefit from strong support networks. Yet, Hannah is a rather passive person who does not believe that physiotherapy can make any real difference for her condition. As a result, she often skips her sessions and remains largely inactive, despite medical advice indicating that physical activity constitutes a protective factor capable of improving treatment response and slowing disease progression. Moreover, Hannah tends to become easily discouraged. When faced with the difficulties imposed by the illness, she quickly abandons sessions, telling herself she will resume it once the situation worsens.

Like Hannah, Mark has generally approached challenges with a degree of passivity. However, he has great trust in his doctors, and for this reason, he decides to adhere consistently to the proposed care plan. He takes his medication daily, participates in regular physiotherapy and physical activity, and gradually begins to notice an increased sense of determination not only in managing his illness, but in life more broadly. Over time, Mark realizes that he has not only become more resolute, but has also developed a deeper appreciation for life and now finds in each challenge a renewed reason to overcome his limitations.

Although it is difficult to predict the health outcomes for either Hannah or Mark as these may depend on a variety of factors, it is nonetheless reasonable to assume that a more perseverant, resilient, and compliant character may be associated with more favorable results. Such a disposition may not only support improved well-being, due to a greater predisposition to face adversity with hope and to recover after setbacks, but may also positively influence the physiological response to illness. It is therefore plausible to imagine that one of these two patients might respond more effectively to the progression of the disease than the other.

The cases of Peter and Mary, as well as those of Hannah and Mark, help illustrate the relevance of a clinical approach grounded in virtue ethics, one that, alongside therapeutic interventions, also fosters patients' work on their own inner world and promotes character education pathways aimed at strengthening individual character strengths and virtues.

This work is clearly not meant to divide persons into type A and type B patients, nor to blame on those who are unable to "engage" in a process of character education. There are situations and contexts in which nothing good can be drawn as nothing appears capable of being turned into something meaningful or positive. This may depend on various factors, such as the severity of one's condition, the presence of severe and uncontrolled pain, the lack of adequate socioeconomic conditions or supportive relational networks. Nonetheless, what I aim to emphasize is that moral flourishing can represent a significant factor in enhancing the quality of life of the ill person. For this reason, it is essential to study the virtues of the patient and to promote educational paths.

3. *Do We Really Need a New Framework?*

In the previous sections, we examined both the strengths and the limitations of care ethics and explored why an approach such as virtue ethics may serve as a valuable complement to it in caring contexts, benefitting both healthcare professionals and patients. However, the strengths of a framework like virtue ethics invite reflection on whether an integrated model is truly needed. Several scholars have argued that virtue ethics is not only self-sufficient, but also capable of addressing the core challenges of care ethics while preserving its most valuable insights (e.g. Halwani, 2003; McLaren, 2001; Slote, 1998a;b). Some thus suggest that it should be subsumed under virtue theory.

Raja Halwani, for example, contends that care ethics cannot subsist as a fully developed moral theory on its own.[4] Integrating it with virtue ethics would allow it to overcome its theoretical limitations without losing its essential contributions such as "its appeal to partiality as a legitimate moral concern, its application to specific individuals in the agent's life, its emotional component, and its relevance to areas in moral life that have been neglected by some traditional moral theories" (Halwani, 2003, p. 162). Thus, Halwani concludes that care ethics should become part of virtue theory, which he considers a more comprehensive and coherent moral framework.

Margaret McLaren similarly supports this position. In her work, she highlights the significant overlap between the two theories, especially in their shared emphasis on relationality and their particularistic moral perspective. Precisely because of these shared features, McLaren argues that virtue ethics is well positioned to absorb and strengthen the core elements of care ethics, while also addressing some of its unresolved difficulties. Accordingly, care ethics should no longer be regarded as an autonomous moral theory, but rather as a subspecies of virtue ethics.

Although care ethics shares several commonalities with virtue ethics such as its emphasis on relational aspects and a particularistic moral outlook, it would be reductive to treat it as a mere component of virtue ethics. Despite its limitations, care ethics offers valuable insights that deserve to be preserved and integrated, especially in care contexts. Reducing the philosophy of care to virtue ethics would mean losing focus on essential themes that other moral theories (including virtue theory) address only marginally, such as the centrality of care. Moreover, this would entail transforming the latter framework with its own distinct characteristics into something it is not as its core lies not in the centrality of care but in the cultivation of character. For this reason, an integrative approach that brings together the two ethical frameworks may offer a more balanced and conceptually robust solution.

Such a perspective is supported by Elizabeth Sander-Staudt (2006). Sander Staudt argues against the marginalization of care ethics and

4 The author focuses in particular on Noddings' account.

contends its importance as it highlights significant dimensions of caring practices often overshadowed by other moral theories, including virtue ethics. According to the philosopher, by placing these elements at the center of moral reflection, care ethics not only reaffirms their importance but also contributes to their conceptual elaboration. Sander Staudt thus advocates for a collaborative framework between care ethics and virtue ethics, proposing that care ethics should enhance virtue ethics as well as be enhanced by it (Stander-Staudt, 2006).

This perspective proves particularly illuminating; rather than reducing or discarding care ethics, Sander Staudt suggests it would be preferable to situate it within a broader framework that draws together the strengths of both traditions. As previously discussed, care ethics would enable both caregivers and cared-for to understand the fundamental role that care plays in human life. This is particularly important because, as Alasdair MacIntyre emphasizes, individuals can remain meaningfully connected to others only if they acknowledge their own nature as fundamentally animal, that is, fragile, vulnerable, and inherently dependent. This is why he emphasizes that "it is important to remember that there is a scale of disability on which we all find ourselves" (MacIntyre, 2001, p. 73). Recognizing our condition of vulnerability and non-self-sufficiency allows us to accept our interdependence, to appreciate what those who are ill can teach us, and to understand the importance of respecting those who find themselves in situations that may one day be our own. By addressing these dimensions of moral life, care ethics may support the development of genuinely patient-centered care models.

Equally important is a theoretical approach that can support professionals in learning how to place the other at the center of moral concern, to understand and respond to their needs (without neglecting their own), to make ethically sound decisions, and to regulate their emotions. Similarly, patients may benefit from an ethical model that promotes the development of virtues supporting both their relationship with caregivers and their ability to cope with illness and its challenges.

In this light, virtue care ethics can open up the possibility of a coherent integration between care ethics and virtue ethics. Such integration recognizes the core importance of care in human life, while also emphasizing the moral significance of relationships and the necessity of being attentive and responsive to the needs of both oneself and others. At the same time, it acknowledges that this can only be achieved through the ongoing cultivation of character. It further underlines the relevance of the agent's moral development not only for enhancing the quality of care provided, but also as a key resource in the experience of illness and in navigating its challenges effectively. In this perspective, virtue care ethics suggests that in order to adequately care for others, to allow others to care for us, and to develop the tools necessary to better manage the context of illness, character work is essential. Conversely, for this to be possible, it is crucial to recognize the moral centrality of care for every human being.

4. *Virtue Care Ethics and "Humanistic" Hospitals*

Earlier sections have illustrated the benefits of integrating care ethics and virtue ethics within a unified framework in care contexts. The discussion has concentrated mainly on the implications of this model for practitioners and patients. However, virtue care ethics may also serve as a valuable resource for transforming care environments and fostering the development of humanistic hospitals (Ricci, 2024a). This model may contribute meaningfully for delivering care that is compassionate and ethically grounded. In fact, facilities should be designed to accommodate patients at different stages of their illness (e.g. Chapman, 2006).

It is worth noting that these elements were already identified as essential by Cicely Saunders in the 1960s. Saunders emphasized the importance of providing what she called "total care", supported by a multidisciplinary team. She also insisted on the importance of creating patient-friendly environments that could genuinely support the well-being of those receiving care. Her vision took shape in the founding of St. Christopher's Hospice in London. To make some examples, Saunders opted for smaller wards composed of rooms with no more than six beds, arranged to allow patients to observe life outside while minimizing health risks. Beds were fitted with casters, enabling patients to be moved easily to the chapel, communal spaces, garden, or even positioned near other patients to facilitate social interaction. Common rooms were equipped with fireplaces and comfortable upright chairs, while the overall design favored colorful, homelike furnishings intended to convey warmth and familiarity (Du Boulay, 1987).

St. Christopher's remains a touchstone for the development of humanistic hospitals and clearly demonstrates the transformative potential of a care model informed by a robust ethical framework. Without such a framework, the clinical environment risks being reduced to a technocratic and procedural space, where the richness of the patient experience is overlooked. In this regard, David R. Graber's critique of modern hospital culture is especially relevant. He notes:

> Modern hospitals are typically large organizations or members of corporations, with a management that is focused on efficiency, productivity, and profit. A "high-tech" workplace, overseen with a corporate, business-centered concept of management are key elements in the culture of today's hospitals. Such a culture is not necessarily deficient or dysfunctional, unless it has supplanted and eliminated the value or ethos of humanistic, compassionate care. In that event, high proportions of patients will experience neglect, be ignored, and not feel respected or satisfied during their visit (Graber, 2009, pp. 517-518).

As Graber points out, most of our hospitals have gradually lost their focus on the patient, shifting their priorities toward values such as efficiency, productivity, and profit. This has resulted in a shift from patient-centered approach toward a "corporate model" of care delivery. However, even large companies like Google no longer adhere strictly to these principles. As literature shows, the better team members are supported and empowered, the more effectively they can contribute to the delivery of high-quality outcomes, thereby improving overall institutional efficiency and productivity (e.g. Tran, 2017).

Moreover, as shown by recent studies, models grounded in compassionate care are not only the most appropriate for patients but also prove to be advantageous for healthcare institutions themselves, contributing to improved organizational performance, greater staff retention, and overall better clinical outcomes[5].

A model such as the one proposed here may appear as impossible to achieve. However, the literature accounts for several examples of humanistic hospitals, starting with St. Christopher Hospice[6]. The commitment to the humanization of healthcare environment is urgent today, as the literature shows that every element may plays a significant role[7]. Not only does the caregiving approach adopted by professionals affect patients' health outcomes and well-being, but structural and environmental aspects can also exert a profound impact. Consider, for instance, studies indicating that proximity to a window can influence the length of hospital stays.

It is evident that in a context marked by limited resources, it may not be possible to implement every potentially beneficial intervention, especially those requiring substantial financial and human capital. Nevertheless, this reality invites reflection on what should be prioritized and what can be improved within the constraints of existing resources.

I do not intend to question the importance of organizational and bureaucratic structures, which are indeed crucial to the functioning of hospitals. However, it is essential to keep the focus on patients, their families, and their needs. To ensure that our facilities are truly patient-oriented, it is necessary to invest in the ethical training of those who design and govern such institutions. Leaders play a decisive role in shaping the ethical climate of healthcare organizations, as their values directly influence institutional priorities, structures, and everyday practices. For this reason, it is imperative to develop training programs that help them recognize the centrality of care and engage in character development, thereby fostering greater

5 See Ricci (2024a) for a more in-depth analysis.
6 For an overview of models of humanistic hospitals see Graber, 2009.
7 On the importance of design in fostering a better inpatient pathway see among others Bates, 2018.

awareness of the ethical impact of their decisions and actions. This kind of ethical formation should be regarded as a prerequisite for any meaningful structural transformation (Ricci, 2024a).

Having outlined an appropriate ethical framework for care settings, it is now necessary to examine whether such a framework can be effectively applied to the patient's experience. This raises the question of whether flourishing can be meaningfully pursued within the constraints and challenges imposed by illness. While it is true that our character influences how we respond to life circumstances, and that a virtuous character can serve as a protective factor supporting better coping and greater acceptance, it is also true that not every context permits the kind of character work needed for flourishing. Therefore, it becomes essential to investigate whether a genuine compatibility exists between the experience of illness and the possibility of flourishing.

PART TWO
ILLNESS AND VIRTUE: TWO MUTUALLY EXCLUSIVE CONCEPTS?

Although human beings spend a considerable part of their lives suffering, being ill, or disabled[1], the moral flourishing and good life of the ill person have received little attention among philosophers. Research (e.g. Bain et al., 2019; Brady, 2018; Carel, 2007; 2016a; b; 2019; Fowers, 2017; Kidd, 2012; 2015; Miles, 2019; Navarini, 2020; 2022; Paul, 2014; Wright et al., 2020a; 2021) has been mainly focused on the character traits which can help the patient's relationship with healthcare professionals[2], as in Edmund Pellegrino and David Thomasma's model (1993). Jonathan K. Miles has singled out two possible reasons for this omission: (i) first of all, since healthcare professionals are the main readers of bioethics literature, to focus on the traits required by providers (physicians, nurses, psychologists, etc.) is perhaps expected to be a priority; (ii) additionally, it is a concern that working on the patient's virtues could be problematic,

1 As Alasdair MacIntyre writes in his *Dependent Rational Animals. Why Human Beings Need the Virtues*, vulnerability, suffering, dependence, disease, and disability are characteristics we cannot ignore, as they derive from our nature (MacIntyre, 1999). Thomas Aquinas also spoke of the typically human element of *passio animalis*, which leads to the emergence of two dimentions: fragility and limitation. See Aquinas, *Quaestiones disputatae de veritate*, q. 26, a. 2. Havi Carel and Kidd also describe human nature as characterized by three "deep features" (Carel & Kidd, 2020) which are *existentialia* for human beings: *contingency, vulnerability,* and *subjection.*

2 It is fundamental to note that being a patient and being an ill person are not the same thing. Being a patient represents one of the "roles" of the ill subject; However, they are not just a patient. By contrast, being ill is a pervasive condition that extends far beyond the clinical setting. It affects the entirety of the person's existence, permeating every domain of life and influencing all the roles the individual inhabits.

as this may cause healthcare professionals to be less caring towards non-virtuous patients (Miles, 2019). However, Miles' intuitions may not be the only reasons for this lack in the literature.

As Ian James Kidd emphasized, there is a fear that moral development may become too heavy a burden for patients, who already have to bear with the effects of a severe illness (Kidd, 2012). As a consequence, what often happens is that ill people end up being reduced to mere "objects of care" and "objects of benevolence" (Navarini & Ricci, 2021, p. 411). Researchers have in fact dwelt on those character traits that can favor patient-operator relationship and compliance, such as docility and patience (see Navarini & Ricci, 2021, p. 651). Although these aspects are undoubtedly important, they encompass only part of the experience of being ill. The ill person is first and foremost a moral subject, and they can therefore be the protagonist of their flourishing process until the very end.

This omission may also be related to the fact that conditions such as death, dying, suffering and being ill have become taboo in our society (Ariès, 1975) and to the stigma attached to many chronic degenerative diseases such as cancer. On the one hand, certain phenomena cannot be talked about; on the other hand, we may be inclined to believe that illness and suffering are somehow "deserved" (e.g. "I deserve" to have lung cancer because I am a smoker). Thus, even the negative consequences of being ill such as suffering appear justified. And on the contrary, positive alternatives to illness, such as living a "normal", full, and "happy life"[3] may be experienced as something precluded from the subject or undeserved.

Lastly, this tendency may run on a misconception, namely, the idea that the subject's overall well-being depends on some conditions deemed essential such as "freedom, health, access to social goods, self-fulfillment" (Carel, 2007, p. 95). If this is true, then a moral flourishing seems to be completely precluded from ill people, especially those suffering from chronically debilitating

---

3 Although "happy" is a controversial term, I choose to use it with reference to Carel's paper *Can I Be Ill and Happy.* In this work the philosopher shows that happiness and a good life are compatible with the experience of illness.

degenerative diseases[4], such as multiple kinds of neoplasms, which often cause permanent disabilities. Even the World Health Organization (WHO) endorses this conception of health, defining it as "a state of *complete* physical, mental, and social well-being and not merely the absence of disease or infirmity" (World Health Organization, 2018).

If we were to take into account all the reasons mentioned above, it would seem obvious that the achievement of a good and happy life cannot be an option for some categories of people, including ill people. However, as Claudia Navarini affirms, "if life can be good, it must be able to be so with pain, with suffering, and with death, from which it cannot be dissociated" (Navarini, 2020, pp. 26-27, my translation). Therefore, whoever wants to delineate an account of a fully accomplished good life cannot fail to consider the potential decline of any subject's condition (Carel, 2007).

In the last few years, the literature has finally begun to move in this direction; philosophers and psychologists have started to investigate the relationship between illness, suffering and virtue from different perspectives. Researchers have thus recognized that being ill does not prevent patients from flourishing and living a good life. Conversely, although it can be a real trauma, it may also hide spaces for a profound transformation. Recent literature has identified increased life appreciation, spiritual change, improved relationships with others, and the development of character traits and virtues as possible consequences of positive adaptation to illness. (e.g. Campbell & Swift, 2002; Cordova et al., 2001; Mystakidou et al. 2008; Park et al., 2009)[5]. The aim of the second part of this book,

---

4 Chronic illnesses are defined by The Centers for Disease Control and Prevention (CDC) as: "noncommunicable illnesses that are prolonged in duration, do not resolve spontaneously, and are rarely cured completely" (Centers for Disease Prevention, 2009).

5 Another interesting work on the positive changes of illness is that of Sodergren & Hyland (2000). The authors have also designed a self-report questionnaire, the Silver Lining Questionnaire, which measures a wide variety of positive changes associated with illness. The items aim at investigating individuals' perceptions of themselves, their relationships with others, their sense of meaning, and their appreciation of life.

which follows the path of the latest literature, is to examine the relationship between suffering, illness, and virtues, demonstrating that moral flourishing and the good life are concrete possibilities for the ill person.

CHAPTER 3
SUFFERING AND PAIN IN THE EXPERIENCE OF ILLNESS

To discuss the relationship between illness and virtue, it is necessary to start with a discussion of the phenomenon of suffering. Indeed, a first and undeniable fact is that dealing with a progressive chronic illness involves suffering. However, to address suffering, it is essential to distinguish it from the experience of pain; although pain[1] and suffering often coexist when one is ill, they do not constitute the same experience[2]. To elucidate this distinction, I will provide an examination of the topic, beginning with Michael Brady's account.

1. *Phenomenology of Pain*

Suffering and pain are often used as synonyms and present similar generative mechanisms. However, while being of the same general kind (Brady, 2018), they refer to two different realms. For example,

1 For an account of the physiology and etiology of pain on which I will not focus here see Giordano & Schatman, 2008 a; b; c.

2 In this work I will often speak, by way of simplification, of pain and suffering as experiences. However, it should be noted that not all philosophers agree in systematically defining these phenomena as such since the subject is not always aware of them. Felice Cimatti contends that "there is no experience unless there is a subject experiencing it, and unless there is something that the subject is experiencing" (Cimatti, 2006, p. 193, my translation). In this sense, in the case of pain asymbolia – a rare condition of the cortex which prevents the suffering subject from perceiving themself as suffering – the subject in no way affected by the events, we should not speak of experience, but of behavior, or of "experiences without subject" (Cimatti, 2012, p. 199, my translation). On this see also Navarini, 2020.

we may say that our heart gravely aches when a boyfriend or a girlfriend breaks up with us; our heart is obviously not physically *harmed* in these kinds of instances, but we nonetheless *suffer*. Put differently, pain refers more squarely to the bio-physical dimension of human beings, whereas suffering is an experience which acts at a deeper level and can engage human beings on various levels. For example, we could say that we have a stomach-ache because we are hungry, since we have not had dinner yet. This, however, differs from *suffering* from hunger. In fact, painful experiences[3], which affect human and non-human animals alike, are often due to the activation of nociceptors[4]. Nociceptors are peripheral receptors which, through painful sensation[5], activate a defense mechanism that warn the subject that they are in a potentially harmful situation. For example, if we touch a flame with our hands, nociceptors will be activated to make us feel pain; as a result, after a short while, we

3 There are also "non-experiential uses of 'suffer' and 'suffering'" (Brady, 2018, p. 11) which fall outside the scope of this work and will not be addressed here. This expression refers to something that is not experienced as negative for its own sake or to cases in which the subject suffers but is not aware of it. An example in this sense could be that of "people who are experiencing health threatening conditions such as lack of education, poverty, unhygienic living conditions, addictions of various kinds, poor diet, dangerous working conditions, and so forth" (van Hooft, 1998, p. 130). To explain what this is all about, Brady in his book gives the example of a football team losing a game or someone suffering from memory loss. This experience, as the philosopher (Brady, 2018) points out, becomes personal when the subject feels it as a reason for displeasure or suffering. The same term can also be used in reference to objects, contexts and situations that undergo negative changes, e.g. water getting contaminated. In this work I will refer to contexts and situations in which the subject feels suffering in relation to facts, events, and conditions, and is aware of them.

4 The term "nociception" stands for "the capacity to sense noxious stimuli" (Allen, 2004) and is one of the most primitive sensory systems that allows animals to avoid potential harmful situations (Tracey, 2017).

5 It is crucial to stress that nociceptive and pain systems do not always correspond. Indeed, firstly, nociceptors do not only respond to pain (e.g. also to heat and touch); secondly, because their activation does not always generate the painful experience (see Brady, 2018, p. 70).

will be stimulated to remove it, thus avoiding possible damage to our tissues.

Another aspect that may enhance our understanding of the nature of pain and distinguish it from ordinary sensory experiences is that pain possesses, as Brady defines it, "an affective-motivational dimension" (Brady, 2018, p. 19) that ordinary sensory experiences lack. To fully comprehend what Brady means by this, let us think for a moment about a persistent pain such as that caused by the complications that may follow the removal of a wisdom tooth. This form of pain will in most cases push the subject to take a painkiller to alleviate their discomfort; if that is not enough, it may motivate them to seek professional dental care. Consequently, for a sensation to be classified as pain, it must possess certain specific characteristics. Pain, indeed, even when it is "functional" or intentionally sought after (e.g. pain caused by physical activity), is (i) accompanied by the *motivational force to remove it*. This is one of the reasons why surpassing one's physical limitations (e.g. for an athletic competition) is particularly challenging: inasmuch as the athlete is required to persist in an action that they would naturally tend to avoid. This motivational force is one of the reasons why pain is different from common sensory experiences.

Other aspects that according to Brady are absent from ordinary sensory experiences but characterize painful experiences (unless the subject lacks motivation to act due to psychological or physiological problems, e.g. severe depression), include: (ii) *cognitive changes*[6] and (iii) *facial* and *expressions changes*.

It is important to note that while the paradigmatic painful experience typically exhibits all of the aforementioned elements, the absence of any single element does not preclude the possibility that the subject is experiencing pain. In fact, while one generally tends to give in to expressions of pain when facing episodes of a certain intensity, it is also possible to control such expressions. This may happen as a consequence of a shock, or because the subject needs to exercise a certain control in a context in which showing

---

6 In particular, the philosopher focuses on the importance of the aspect of attention. See also Ecclestone & Crombez, 1999.

fragility could constitute a danger. Let's think, for example, about the eighth episode of the Netflix series *Squid Game*, season one, "Front Man". In this episode, it is shown how Kang Sae-Byeok tries to avoid showing expressions of pain during dinner, to appear strong in front of her rivals. The wound caused by the glass that was stuck in the girl's belly must have made her grimace; moreover, the significant blood loss had reduced her energy. However, controlling the expressions of pain and fear was essential to Kang Sae-Byeok to survive the ruthlessness of the other players.

In conclusion, in response to the inquiry regarding the nature of pain, it can be posited that pain is a physiological mechanism activated by the stimulation of nociceptors. The experience of pain is invariably accompanied by a potent motivational drive to remove it; cognitive and somatic alterations are frequently observed.

2. *What is Suffering?*

Having delineated the essential components that characterize the experience of pain, we shall now proceed to elucidate the concept of suffering.

The term suffering generally refers to a negative experiential state[7] and encompasses a psychological component. This psychological dimension is often linked to an emotional response to pain, distress, or hardship. Unlike pain, which can be more easily localized and measured through physiological responses, suffering is broader, involving cognitive and affective elements that contribute to the subject's perception of their condition.

In the context of illness, suffering can manifest in multiple forms and affect the afflicted individual at various levels[8] to the extent

---

7 Even if we can define it as a negative experiential state, there are some cases in which a subject derives a sort of pleasure or satisfaction in suffering (e.g. victimhood).

8 It is fundamental to underline that the different levels of human existence are connected. In fact, profound physical suffering may compromise the patient's mental state, just as mental suffering may worsen the physical pain experienced by the patient. A telling example in this regard is

of becoming entirely pervasive. It is precisely for this reason that Cicely Saunders, the founder of modern palliative care, named this latter form of suffering "total pain"[9].

Looking at the experience of suffering more broadly, according to Brady, the primary characteristic of suffering can be identified as a negative state, marked by an *unpleasant feeling*[10], which is

---

recounted by Cicely Saunders. During her time at St. Joseph's Hospice. A woman explained to Saunders that her pain had begun in her back but gradually came to affect every part of her being. In one account, the woman described how her pain, initially located in her back, gradually extended to affect every aspect of her being. Saunders wrote about this: "One person gave me more or less the following answer when I asked her a question about her pain. [...] She said, "Well doctor, the pain began in my back, but now it seems that all of me is wrong". She gave a description of various symptoms and ills and then went on to say, "My husband and son were marvelous, but they were at work, and they would have had to stay off and lose their money. I could have cried for the pills and injections although I knew I shouldn't. Everything seemed to be against me, and nobody seemed to understand" (Saunders, 1964, p. viii). This can help understand more thoroughly that the different levels of suffering are all profoundly connected. They have the power to transform one into another: psychic suffering can be the cause of physical pain, and vice versa. In some cases, suffering can even become existential. For this reason, it is crucial to approach care holistically, addressing all patient's needs.

9 Although "total pain" is a conventional term in palliative care literature, this concept may be misleading. It is therefore essential to clarify that Saunders used the term "total" to refer to a type of pervasive suffering which is not limited to the physical sphere but permeates all levels of the ill person. On total pain see also Clark, 1999.

10 There are two main schools of thought trying to explain the nature of this *unpleasant feeling*: the "internalists " and the "externalists". The internalist perspective (also called the "phenomenological" or "felt quality" view) holds that unpleasantness is a feeling internal to the unpleasant experience. This view is divided into the "distinctive feeling view", and the "hedonic tone view". For the distinctive feeling theory, an experience is unpleasant if it contains "a distinctive kind of feeling, one may call ... 'the unpleasant feeling'. [...] The Hedonic Tone Theory, on the other hand, maintains that the feeling of unpleasantness is akin to an aspect of some distinctive feeling, rather than being identical with a distinctive feeling itself" (Brady, 2018, p. 33). This view has over time generated many perplexities because if it were true that the feeling of unpleasantness was internal to the

also accompanied by other significant elements that distinguish it from other negative states (including pain), namely *intensity* and *importance*. As Brady writes, "one suffers when they experience a negative affect of a suitably intense kind. On this view, it is only intense pain or bitter cold, for example, that constitute kinds of suffering" (Brady, 2018, p. 23). Furthermore, he adds: "The object of one's experience must be important or significant enough for one's unpleasant experience to count as suffering. On this view, one suffers only when one's social standing is significantly threatened, for instance, such that one is mortified with shame or embarrassment" (Brady, 2018, p. 23).

So, to say that a subject suffers, their experience must be negative, intense, and hold profound importance for the person. For example, an individual may experience intense discomfort (in some cases even pain) due to exposure to low temperatures without adequate protection. However, this situation would only become a real cause of suffering if, say, they were homeless, constantly exposed to the harsh Swedish winter, and had no possibility of sheltering anywhere during the night. In such circumstances, the sensation of discomfort (or pain) would be transformed, reaching maximum intensity,

experience itself, it would be complicated to describe it. There are several criticisms that have been made to this theory, the most significant being the "heterogeneity problem", which holds that since we can distinguish so many different types of unpleasant experiences it becomes really complicated to understand what exactly a distinctive feeling or a hedonic tone have in common. The second proposal, the externalist view, maintains that the unpleasantness of such states is related to external elements, and that unpleasant experiences do not have a particular kind or feeling in common, they rather share a *similar attitude*. There are several externalist approaches: the "desire theory", the "evaluativist theory", and the "imperativist theory", depending on which attitude is believed to be the unifying element of the different unpleasant experiences. These theories are referred to as externalist because they do not consider the feeling of unpleasantness as internal to the experience itself. Instead, it is regarded as "a property or quality that we ascribe to the sensation in virtue of its relation to an external factor or element" (Brady, 2018, p. 47). The most obvious reason why it is easier to lean towards these theories is that they do not have to face the heterogeneity objection since they defend the diversity of characteristics with which unpleasantness can present itself.

importance, and a strong motivation to alleviate the situation, even if resolution were impossible. In this scenario, one could assert that the individual is suffering.

To better understand the difference between pain/discomfort and suffering, I quote below an example given by Brady: "My dull pangs of hunger might constitute a state of suffering because I'm desperate and living below the poverty line and *always* feeling this way, and this fact distinguished my low-intensity hunger from similar low-intensity hunger that is a precursor to a rich person's midday snack" (Brady, 2018, p. 24).

In this sense, it could be said that suffering encompasses a psychological component that cannot be straightforwardly reduced to bodily mechanisms. This implies that only certain types of pain cause a subject to suffer. Intense and persistent pain may carry significance for the person experiencing it, and consequently, it may become a source of suffering. It is precisely in this sense that Brady argues that pain is one of the "building blocks" (Brady, 2018, p. 23) of suffering. However, as distinct phenomena, pain and suffering do not invariably co-occur. In fact, there are sufferings which do not involve the subject's experience of physical pain, and some pains which do not generate any kind of suffering. Examples in this latter sense are supplied by cases of "pain asymbolia", a rare condition caused by a lesion affecting the posterior insula which prevents the pain-feeling subject from perceiving any kind of suffering, as they remain completely indifferent to the pain experienced (Klein, 2015, p. 493). Additional instances of circumstances in which an individual may experience pain without suffering include the sensations of tiredness, hunger, cold, irritation, and nausea. These conditions may cause pain but, in most cases, they are not of sufficient intensity to constitute episodes of suffering. This distinction does not imply that the subject never experiences suffering in such contexts – prolonged hunger can indeed become a form of suffering – but it elucidates the distinction between pain and suffering, demonstrating that they are not necessarily interconnected. For example, certain forms of pain can elicit pleasant sensations. In this sense, let's think for example to engaging in activities such as training with substantial weights or participating in a marathon. In both cases, the subject is expected to experience a degree of physical discomfort; nevertheless, these activities represent essential conditions for achieving outcomes that athletes find deeply gratifying and rewarding. In the same way, eating a bag of Cheetos implies a burning sensation, which may even lead the subject to experience a mild form of pain; however, Cheetos lovers would not characterize the experience as causing any form of suffering.

Likewise, suffering can also exist without pain. For example, a patient who uses analgesics may not feel pain, nonetheless they may suffer psychologically, morally, socially, spiritually, and so on. While the nature of pain should now be apparent, it is equally evident that pain is primarily associated with physical sensations. In contrast, suffering invariably encompasses a mental component.

The last important aspect that characterizes suffering, as Brady has pointed out, is the *attitude towards negative situations*. Suffering is predominantly directed toward the meaning that a given situation conveys. This is particularly evident when considering illnesses with an unfavorable prognosis. For example, a patient with end-stage metastatic cancer, whose pain is managed through appropriate therapy, will likely not experience pain, although they may suffer for what that illness means to them and for what it implies. The fear of not being able to see the smile of the loved ones again, the knowledge that death will inevitably make them suffer, the anxiety of not being able to manage one's financial affairs, and the distress of not having sufficient time to resolve personal or spiritual matters[11] are only some examples of circumstances and thoughts regarding the future that may cause the patient to suffer.

It is for this reason that the recourse to palliative care is fundamental. Palliative care has been defined by the International Association for Hospice and Palliative Care (IAHPC) as "the active holistic care of individuals across all ages with serious health-related suffering due to severe illness, and especially of those near the end of life. It aims to improve the quality of life of patients, their families and their caregivers" (IAHPC, 2018). This approach aims

11 This does not mean that the object of suffering must be intentional; as Brady points out, an object does not always exist. Consider, for example, the phenomenon of phantom limb pain, which can cause genuine suffering despite the absence of the limb itself.

at restoring the overall well-being of patients (even when a complete remission is no longer possible) and to promote their overall well-being, while also considering the needs of family members and caregivers. The fundamental importance of these aspects will be addressed subsequently.

Early palliative care not only improves patients' quality of life but also reduces the incidence of depressive symptoms and the use of disproportionate, aggressive treatments in the final stages of illness. Moreover, it helps patients feel supported rather than abandoned. Finally, early palliative care has often been associated with increased life expectancy (e.g. Corsi et al., 2019; Ferrell et al., 2017).

If, as this evidence suggests, patient adherence to palliative care is crucial, it becomes a collective responsibility to promote its wider diffusion and to regulate its application, particularly in its early and simultaneous forms, delivered alongside first-line treatments, so that it may become a concrete, accessible option for all patients.

Having addressed the previous considerations, we can now come back to suffering *per se*. If we were to provide a definition of suffering[12] that would take into account all of the aforementioned aspects, we could say, using Brady's definition, that "a subject suffers when and only when they have an unpleasant[13] or negative affective

12 Brady's proposal has been recently attacked by Antti Kauppinen, according to whom the element shared by various unpleasant experiences is not the desire to avoid feelings such as suffering. On the contrary, according to Kauppinen it would be a "negative affective construal", i.e. the feeling of unpleasantness experienced leads us to feel a desire for change. The negative affective construal should be composed of three elements: (i) the perception that the situation in which we find ourselves requires a change; (ii) a real desire for such a change and (iii) the belief (or perception) that such a change is not in our power and does not depend on us" (Kauppinen, 2020, p. 20).

13 In this sense Brady embraces a "desire view of suffering". Desire views hold that the feeling of unpleasantness is characterized by two aspects. First, it is a feeling; second, it is always accompanied by the desire that the feeling not be occurring (Brady, 2018, p. 46). To give a definition, the desire theory of unpleasantness proposes that: "[a] sensation S, occurring at time t, is unpleasant at t iff the subject of S desires, intrinsically and de re, at t, of S, that it not be occurring at t" (Heathwood, 2007, p. 41). According to Brady this view has a problem: most theorists, in fact,

experience that they mind, where to mind some state is to have an occurrent desire that it not be occurring" (Brady, 2018, p. 27)[14].

Although this conceptualization appears relatively robust, it seems to lack certain aspects that Brady himself considers as fundamental and that are essential to distinguish the experience of suffering from other situations involving a sense of unpleasantness. For this reason, I propose a minor transformation of Brady's definition into what I will refer to as the "new definition of suffering":

> *New definition of suffering*: a subject suffers when and only when they have an unpleasant or negative affective experience that they perceive as deeply intense, which they therefore want to get rid of, and of which they mind; where to mind some state is to have an occurrent desire that it not be occurring, in virtue of the importance and meaning it has for the subject.

In summary, suffering can be primarily conceptualized as a mental or psychological state characterized by a sensation of unpleasantness and the desire for its cessation. It is distinguished from negative states and pain by certain attributes: intensity, significance, and the individual's disposition towards it. As previously discussed, it can arise from various sources: one may experience suffering due to hunger, physical discomfort (e.g. migraines), spiritual distress, interpersonal difficulties, psychological distress, and other factors. Additionally, individuals with chronic or degenerative conditions may encounter a distinct and multifaceted form of suffering, which

---

emphasize the importance of the aspect of sensation, while for the philosopher unpleasantness is rather a feature of the experience itself. Therefore, he proposes a modification with respect to Heathwood's proposal. Brady's *New Desire Theory* holds that: "[a]n experience E, occurring at time t, is unpleasant at t if and because E consists of a sensation S at t, and the subject of S desires, intrinsically and de re, of S, that it not be occurring at t" (Brady, 2018, p. 48). Brady's view seems to better take into account what happens in real life than Heathwood's. In fact, as Brady points out, what we want to stop when we are suffering is the experience itself, and not just the feeling.

14 By the minding aspect included within his definition, Brady means the importance that such experiences have for the subject; we do not, in fact, become concerned with what lacks importance to us.

can be classified as existential (or total) when it affects the afflicted person across multiple dimensions.

To elucidate the distinctions between pain and suffering, a summary table is presented below.

	Pain	Suffering
Description	It refers to the bio-physical dimension of human beings (nociception).	It is "internal" and therefore has a mental component.
Characteristics	Affective motivational dimension; may involve changes (e.g. attention, expressions).	Affective motivational dimension; may involve changes; *deep intensity; profound importance*.
Example	This headache is giving me *pain*, and I would like to get rid of it. However, although I notice *changes in me*, such as struggling to concentrate, the pain is not so intense that I can't continue with my daily activities.	I am *suffering from* migraine; its *intensity* and *impact* are so overwhelming that it is preventing me from doing anything else. All I want is for it to stop.

Tab. 1

3. *The "Short Circuit" of Suffering*

Having examined the primary distinctions between pain and suffering and delineated the characteristics of suffering, it is pertinent to further analyze the concept of total pain, which encompasses the existential suffering that may be experienced, potentially affecting every aspect of a patient's existence. One of the most significant problems which may lead chronically ill people to face total forms of suffering, regards the characteristics of suffering itself: the more profound an individual's suffering, the greater their desire to alleviate that distressing experience. However, one of the main features of chronic and degenerative diseases (especially in advanced or terminal stages) is the impossibility of avoiding them. This may generate a "short circuit": illness induces the subject to experience suffering, yet the inherent nature of suffering tends to elicit a desire to eliminate its source. Nevertheless, the primary characteristic of chronic conditions is precisely the fact that they cannot be eradicated. This may engender a mechanism of frustration that could lead the patient to feel "trapped": if the source of my suffering cannot be eliminated, I have no alternative but to be ill, suffering, and unhappy for the rest of my life.

Nevertheless, upon reflection, every human being undergoes a series of adaptations throughout the course of their life. These adaptations also include learning to live with various forms of suffering. As Havi Carel and Ian James Kidd write:

> Every day we change a bit (or a lot) in the kinds of preferences, goals, and desires we have. Again, otherwise we would not make the transition from childhood to adulthood, and on to older age. We become better planners, with more realistic goals; we become better at recognizing and controlling our desires, we have increasingly mature preferences and values, and so on. Again, at least partly, these changes are the result of experiences that transform us (Carel & Kidd, 2020, p. 11).

Although some changes may be particularly frightening, we should not think of them as "traps". Change may indeed hide positive opportunities, even when it comes from the most brutal circumstances[15]. Experience, surprisingly, supports this assertion: there exist individuals who are ill yet happy or even flourishing by

---

15 As Navarini writes: "suffering and mortality are good in any case, to the point that the phenomena of negation of suffering and death do not make life better, but falsify it, illusorily transforming it into something that is not and cannot be given in the authenticity of experience" (Navarini, 2020, p. 27, my translation).

virtue of such an experience. Illness and suffering, consequently, should not be seen as a threat to the subject's well-being; this threat rather looms in the way the patient responds to their disease.

Pain and suffering are integral components of human existence, serving as natural mechanisms that alert individuals to potential dangers and adverse conditions, both internal and external to the organism. It is precisely due to our predisposition to such phenomena that we are inherently inclined to respond and adapt to them. The type of adaptation, whether positive (e.g. evidenced by the acceptance of new living conditions, or by the development of character strengths and virtues) or negative (e.g. manifested in forms of anxiety, depression, distress, which can become stable conditions), depends on the patient's response to illness and suffering. Each individual's response is unique, reflecting the inherent diversity of human experiences. However, not all approaches in healthcare consider diseases from this perspective. Conversely, there is a tendency in medicine to regard illness as a worthless bodily affliction.

4. *Why Illness Demands More Than a Naturalistic Lens*

There are, in fact, approaches such as the "naturalistic" one which tend to merely describe illnesses as a dysfunction of the body, excluding the first-person experience (see Boorse, 1997). However, the physiological issue is only one of the aspects involved, and as Carel (2007) convincingly argued it says nothing about the experience itself. This latter reason indicates the importance of complementing the naturalistic perspective with a phenomenological framework, which focuses more on illness as experienced by those who live it rather than just on the physiological dysfunctional state itself. Let us now delve into the phenomenological view of the experience of illness.

Phenomenology is a philosophical current which focuses on how experiences (phenomena) are perceived and lived by the subject, rather than on their nature. While phenomenology is highly attentive to objective empirical data, it does not absolutize such information. Instead, it considers the objective and subjective components of experience as possessing equal ontological dignity. This approach is particularly significant in such contexts, as the objective characteristics of illness, albeit fundamental, constitute only a small component of the individual's experience. An illustrative example that elucidates the importance of a mixed approach can be derived from Carel's book *Illness. The Cry of the Flash.* In her book, Carel shows that a naturalistic approach can provide some factual information about an illness in general; however, it does not say much about what illness means to the patient experiencing it. Taking depression as an example, the naturalistic perspective would not account for "the qualitative texture of waking up depressed, the dark thoughts, the listlessness, the sense of doom, and so on" (Carel, 2019, p. 10). Nevertheless, these aspects are fundamental to the subjective experience of the individual, influencing how they perceive and navigate their condition. Explaining depression from a naturalistic point of view may be sufficient to comprehend the phenomenon in general; however, it is inadequate for describing the subject's personal experience or determining the appropriate treatment for that patient. Indeed, given that each patient is unique, their individual suffering will also be distinct[16].

---

16 Another philosopher who argues in favor of this approach is Karen Lebacqz. Lebacqz writes on this topic: "The physician may know *what* the patient has, but only the patient knows what it *means* to have it. Illness is not merely a physical phenomenon, with symbolic meaning. The symbolic meaning of illness is not totally idiosyncratic, since it will be formed in large part by cultural expectations. But it is nonetheless personal. Only the patient knows this personal side of illness" (Lebacqz, 1992, p. 302). According to the philosopher, the virtue of humility would play an important role in regard to this as it (i) would allow professionals to recognize that the patient knows more about what they are experiencing than the professionals themselves. Also, (ii) it would create room for a better understanding of the meaning and nature of the disease, and of the way in which it manifests itself to the subject (see Lebacqz, 1992, p. 302). In this sense, such a virtue is a resource for health-care professionals because it would enable them to better understand not only the disease but also the illness as experienced and perceived by the patient.

In the last few years, several scholars have explored the relationship between phenomenology[17] and illness (e.g. Carel, 2011; Kirkengen, 2007; Svenaeus, 2000a; 2000b; Toombs, 2001; Zaner, 2005) focusing on the centrality of the body in human existence. According to their perspective, a body can be both an "objective body", and a "lived body". Carel delineates the difference between objective body and lived body[18] as follows: "This is the difference between the biological body and the lived body. The biological body is the physical or material body – the body as object. This body can become diseased. The lived body is the first-person experience of the biological body. It is the body as lived by the person. This body experiences illness" (Carel, 2019, p. 33).

This approach evidently draws significant influence from Maurice Merleau-Ponty, who posits that the body is for human beings "the origin of the rest, expressive movement itself, that which causes them to begin to exist as things, under our hand and eyes" (Merleau-Ponty, 2013, p. 146). Regarding the relationship between the lived body and the objective body, Merleau-Ponty argues that when an illness occurs to the subject's life, the differences between the objective body and the lived body tend to be accentuated (Carel, 2011). In everyday life we tend to perceive our body as a unified entity: for example, when seated on a sofa, a person does not discern any distinction between one's objective body and lived body. Conversely, during illness, we begin to focus attention on what does not work, resulting in the dimension of the lived body becoming predominant. Elena Colombetti write on this topic: "our touch opens us up to external reality, we feel hot or cold, we perceive a caress, the smoothness or

17 The naturalistic approach and phenomenology are not the only accounts which can be used to talk about illness. Normativism offers a different perspective by viewing illness as value-laden. This account maintains that in order to understand illness and disease, it is necessary to examine how societies perceive and define the condition of being ill. One of the main limitations of this approach lies in the absence of the first-person perspective, which remains essential for grasping the lived experience of illness.

18 This definition also helps to understand the differences between illness and disease.

roughness of an object; but if we have a wound, any contact with that part of the body makes us perceive not so much the object but the aching body" (Colombetti, 2009, p. 18, my translation).

In the experience of illness, aspects of our lives that we previously took for granted thus become bearers of meaning. As Descartes argued, illness removes the "bodily transparency", and compels the subject to problematize it:

> There is nothing that my own nature teaches me more vividly than that I have a body, and that when I feel pain there is something wrong with the body, that when I am hungry or thirsty the body needs food and drink, and so on… Nature also teaches me, by these sensations of pain, hunger, thirst and so on, that I am not merely present in my body as a sailor is present in a ship, but that I am very closely joined and, as it were, intermingled with it, so that I and my body form a unit. If this were not so, I, who am nothing but a thinking thing, would not feel pain when the body was hurt, but would perceive the damage purely by the intellect, just as a sailor perceives by sight that his ship is broken (Descartes, Meditations on First Philosophy, §VI, AT 81).

Thus, when a disease occurs, it affects not only the body as an object, but it also transforms the lived body as experienced by the subject, thereby altering one's perception of the world, their way of being, their self-perception, and so on. This further helps us to grasp why a naturalist approach is not enough to describe the experience of being ill or of taking care of a patient.

Karen Lebacqz further explores this phenomenological perspective on illness. She describes it as an "unwelcome intruder" which brings suffering and pain to the ill subject's life. Lebacqz observes that when someone contracts an illness, it provokes the need for a redefinition of the image that a patient has of themselves (see Lebacqz, 1985, p. 277). This redefinition requires adaptation to the new conditions and circumstances; in a sense, one may posit that life essentially recommences following the diagnosis of a severe condition. Reading what Carel relates about her own illness, it is possible to gain insight into the profound alterations required of individuals facing health challenges. Carel states: "Everything I had until that point taken for granted was thrown into the air. I experienced extreme uncertainty and anxiety. I had to adjust to, and make sense of, a huge shift in my life plans and expectations and my day-to-day routine" (Carel, 2019, p. xi).

Upon receiving a diagnosis (and the more severe the diagnosis is, the more intense this experience is for the patient) the subject's physical, psychological, relational, and social worlds may be extensively altered. Consider the case of a promising young fencer, Carl, who has been diagnosed with osteosarcoma, and must undergo an amputation to minimize the chances of cancer recurrence. Carl will need to work on his self-acceptance to adapt to his altered physical appearance. Additionally, he will need to commence physiotherapy sessions, demonstrating consistency, patience, and perseverance to optimize his recovery. Furthermore, he will need to start a psychological journey to reassess his plans and confront the various challenges that the disease will present.

This example illustrates the multitude of changes that illness introduces, and the substantial effort required from the affected individual. This new beginning demands creative measures to manage daily life and adapt to new circumstances and limitations. However, this process unfolds through a journey marked by various emotions and distinct stages. I will now examine the phases typically experienced by individuals in their journey, drawing on the most influential account of the stages of grief, namely, that offered by psychiatrist Elizabeth Kübler-Ross.

5. *The Stages of Grief and Adaptation*

Kübler-Ross's stages of grief have been extensively studied in the literature and constitute a significant reference point for understanding the ill person's pathway. According to the psychiatrist's account, the patient's journey encompasses five phases: denial (and isolation), anger, bargaining, depression, and acceptance[19]. Hence, the term DABDA is often utilized to define them.

---

19 Even Friedrich Nietzsche, who extensively explored the theme of suffering in his writings, speaks of certain "moments" that the suffering person goes through. According to Nietzsche, the sufferer (a) would first

Following the communication of the diagnosis, the patient initially experiences a state of shock, provoked by the unexpected nature of the news and often accompanied by intense negative emotions such as the fear of dying. What follows is the stage of *denial*. Kübler-Ross wrote on this: "[T]he patient's first reaction may be a temporary state of shock from which he gradually recovers. When his initial feeling of numbness begins to disappear and he can collect himself again, man's usual response is "No, it cannot be me"" (Kübler-Ross, 2009, p. 34). As the state of shock subsides, when patients comprehend the reality

---

experience the feeling of *ressentiment*. Then, (b) they would seek a way to be discharged from the latter, but as they cannot (because the only way would be to turn and vent their anger against a culprit who does not exist), (c) this would lead to a potential "suicidal nihilism". Indeed, the fact that no one is responsible for such suffering is unacceptable to human beings (GM III: 28). Nietzsche writes about this: "What actually arouses indignation over suffering is not the suffering itself, but the senselessness of suffering" (GM II: 7). In fact, it is not the fact of suffering that is unacceptable – also because, according to the philosopher, suffering is a central aspect of human existence – but rather that such suffering is meaningless (i.e. caused by nothing and without a culprit, Leiter, 2003). Thus, the only way to "detonate" the *ressentimenti* caused by the latter would be, according to Nietzsche, the ascetic life. It should be clear by now that, unlike Kübler-Ross, Nietzsche does not believe that it is possible for the subject to achieve any inner peace or acceptance of their own condition, as the only way to escape from suicidal nihilism remains, according to the philosopher, the ascetic life. However, the latter is nevertheless an instrument of expiation of guilt. In fact, according to Nietzsche, if *ressentiment* cannot be vented on another subject (or object), then the sufferer will be punishing themself as guilty. As Brian Leiter writes: "In short, the sufferer himself is to be the object of his own *ressentiment* since he is taught that he himself is the cause of his own suffering. As a result, the sufferer now has "a living being upon whom he can release his emotions" (GM III: 15), namely themself. Only suffering that has a culprit can be bearable" (Leiter, 2003, pp. 260-261). Lastly, according to Nietzsche, suffering should not be escaped, but rather welcomed and experienced, as it is necessary for the development of strength and the achievement of excellence (Brady, 2018, p. 94).

of their illness, the initial feeling of denial is supplanted by emotions such as anger, rage, envy, and resentment[20].

Once the patient comes to terms with the fact that what is happening is indeed happening to them, they often begin to seek reassurance through potential solutions to their condition. This typically marks the beginning of the *bargaining* phase. During this stage, the patient may believe that adhering to a specific diet could lead to a regression of the illness; a child may think that by giving up what they love most, their parent will recover. Others may hope that by undergoing a meaningful experience or sacrificing something of great personal value, they might be spared from the progression of the disease. This occurs because, as Kübler-Ross explains, the individual: "[K]nows, from past experiences, that there is a slim chance that he may be rewarded for good behavior and be granted a wish for special services. His wish is most always an extension of life, followed by the wish for a few days without pain or physical discomfort" (Kübler-Ross, 2009, p. 66).

Bargaining thus represents an attempt to avoid or "postpone" what one fears, and generally includes a prize offered "for good behavior" (e.g. If I recover, I will do more charity), or a self-imposed "deadline" (e.g. I just want to be there at my son's wedding) (Kübler-Ross, 2009). However, as the subject realizes that what they are experiencing is not a temporary condition and that bargaining will not lead to the hoped-for change, the patient will tend to enter the fourth phase: *depression.*

According to Kübler-Ross, there are two kinds of depression that may be experienced in the course of illness: a "reactive depression", and a "preparatory depression". The first one is mainly addressed to the past, and it may occur as a response to a disease. A case of reactive

20 To better explain this state, Kübler-Ross offers an example taken from her own professional experience. During an interview with one of her patients, Dr. G., said: "I suppose most anybody in my position would look at somebody else and say, 'Well, why couldn't it have been him?' and this has crossed my mind several times... An old man whom I have known ever since I was a little kid came down the street. He was eighty-two years old, and he is of no earthly use as far as we mortals can tell. He's rheumatic, he's a cripple, he's dirty, just not the type of a person you would like to be. And the thought hit me strongly, now why couldn't it have been old George instead of me?" (Kübler-Ross, 2009, p. 40).

depression may emerge from the distress the patient experiences upon receiving a diagnosis, from the worsening of their condition, or from the realization that what they once hoped for themselves is no longer attainable. In describing this phase, Kübler-Ross does not refer to major depressive disorder, but rather to a transient emotional state triggered by a specific event. It is worth noting that healthcare professionals, family and caregivers play a very important role in dealing with this phase. As Kübler-Ross explains, it is surprising how quickly a patient's "depression" can be lifted when their needs are taken care of. This, of course, does not exclude the possibility that depressive feelings may evolve into a clinical condition. However, it is also possible that they represent a transient state.

Turning now to the second type of depression identified by Kübler-Ross, preparatory depression is directed instead towards the future, and it is the most typical form of depression at the end of life. It is a tendentially more silent kind of suffering, in which "there is no or little need for words [...]. It is a time when too much interference from visitors who try to cheer [the subject] up hinders his emotional preparation rather than enhances it" (Kübler-Ross, 2009, p. 71). This type of depression, unlike the first one, is difficult to treat with words of consolation since it characterizes the time of preparation for the passing. During this phase, the patient needs time to spend with their loved ones, time to spend alone, time to accept death serenely.

Dealing adequately with the preparatory depression is essential according to Kübler-Ross. It is in fact only those who have dealt adequately with this phase who can enter the last stage of grief, that of *acceptance*. According to the psychiatrist, acceptance is the moment which precedes death. Kübler-Ross holds that acceptance is possible only when the patient has faced and overcome all the previous moments: denial, anger, bargaining, and depression. During this phase, the ill person generally needs more rest and longer hours of sleep. This does not occur because they have given up, but rather because they have entered a state of serenity, almost a void of feelings[21].

21 This is also particularly important, as Navarini (2020, p. 44) has pointed out, to better assess fundamental bioethical issues, avoiding for example the risk of demanding disproportionate treatments.

Acceptance, as Kübler-Ross makes clear, is particularly important when death approaches as it encourages a more careful assessment of end-of-life care, avoiding for example unreasonable and disproportionate treatments. An acceptance of one's condition can help the patient understand that everything possible has been done and that the time has come to say goodbye to their loved ones and to attend to the matters they consider important.

It is worth noting that such acceptance does not occur outside of a process. While there are indeed cases in which patients simply come to terms with their condition and its progression, the opposite can also happen. The person may feel overwhelmed by despair or fear and may consequently request another line of therapy. This often occurs when the patient has not been sufficiently supported or prepared to come to terms with the trajectory of their condition. Moreover, it may particularly happen when the fear of suffering or pain is present. Some patients may experience anxiety about financial concerns, others may suffer deeply from the awareness that they will no longer see their loved ones, while others may fear the uncertainty of what lies ahead. For these reasons, it is crucial to support the ill person during this phase not only by offering reassurance regarding physical care, but above all by ensuring the full spectrum of assistance they may require, whether psychological, moral, social, or spiritual. It is through such a process that patients and their families can be provided with the tools to better accept the various stages of illness, including the preparation for death. This demands presence, attentiveness, the ability to offer the right proximity, and the capacity to respond creatively to the person's emerging needs.

Another fundamental element which, according to Kübler-Ross, is important during end of life is *hope*. Kübler-Ross writes that hope "paradoxically [...] is the more constant companion in the dying experience" (Kübler-Ross, 2009, pp. xiii-ix).

According to the psychiatrist, hope is not a virtue, but rather as a deep feeling (Navarini, 2020a, p. 45) which enables the ill person to react appropriately to the different phases of death and dying[22]. Hope

22 As Kübler-Ross suggests: "No matter what we call it, we found that all our patients maintained a little bit of it and were nourished by it in

is crucial at the end of life, as it can promote acceptance. However, it must not become a desperate attachment to life. Working on the patient's hopes is thus particularly important, since placing hope in unattainable goals may obstruct the possibility to accept.

Kübler-Ross's stages are considered by the literature one of the most authoritative accounts of the internal journey experienced by the patient following the communication of the diagnosis. However, they have not been exempted from criticism. As Navarini points out, they do not tend to capture the entire experience of the dying (or grieving) person, partly due to the assumption of a fixed succession of phases (see Navarini, 2020, p. 44). In fact, terminally ill individuals do not necessarily go through every stage of grief. Moreover, the stages are often not experienced in a linear sequence. This objection has been counter-argued by Kübler-Ross herself, who partially revised her theory, explaining that the moments of grief (i) can be cyclical and (ii) in some cases may not all occur (e.g. Kübler-Ross, 2009; Kübler-Ross et al., 1972; Wessler & Avioli, 1972).

Another point of contention in Kübler-Ross's account is the idea that the phase of acceptance may occur only prior to death. As in the case of depression, two different types of acceptance may be distinguished: a "reactive acceptance", and a "preparatory acceptance". In fact, while it is true that there may be a preparatory form of acceptance preceding death, there may also be (indeed, this is highly desirable) a moment in which the subject accepts their new condition, along with the changes and transformations that illness brings. This kind of acceptance is particularly important because, as has been demonstrated by a large body of literature (e.g. Guzek & Kowalska, 2009; Pankowski et al., 2022), it can help patients to reinterpret their illness as a meaningful part of their life journey and to move forward with it. It is possible that this type of acceptance was not mentioned by the psychiatrist because her work focuses specifically on death and dying. Nonetheless, it is crucial to recognize that it represents a key step in the process, and one that must be actively addressed in the patient's care.

---

especially difficult times" (Kübler-Ross, 2009, p. 113).

As I have already noted in this chapter, accepting and positively adapting to illness should not be taken for granted. It requires the ability to let go of one's previous life, to embrace a new one, and to come to terms with its limitations. Although this is undoubtedly a challenging process, not only is positive adaptation to illness possible, but it is also vital, as it can enhance the patient's overall well-being. This is precisely why the next chapter will focus on adaptation: how it occurs in the context of illness and why supporting patients in this process is essential. In particular, we will examine the relationship between positive adaptation and posttraumatic growth. Posttraumatic growth is a potential (positive) response to trauma that may manifest in beneficial changes to the individual's character. Fostering the conditions for its the development is important because, as we will see in greater detail, it can strengthen the patient's resilience, encourage growth in their moral character, and, more generally, promote a deeper state of well-being.

CHAPTER 4
ACCEPTANCE, COPING, AND POST-TRAUMATIC GROWTH
Illness as a Pathway to Virtuosity

1. *Adaptive Coping: a Difficult Process*

As I pointed out in the previous chapter, pain and suffering are experiences that affect individuals from an early age (See Navarini, 2018) – the fact that life begins with crying is particularly representative in this respect – even though not every moment of life is characterized by them. Pain and suffering may appear unjust, and in some cases as unbearable facts. However, they are strongly embedded in our existence, and our biology gives us the tools to deal with them.

As Navarini specifies, there are several ways through which we can respond to pain and suffering. Human beings can react physically and emotionally to those [B] by escaping, with crying, with shock[1]. We can anticipate pain [A] (e.g. with fear); we can resist it [B.a]; we may try to remove its cause [B.b] (e.g. by taking a drug to calm the pain we feel because of headache); or we can remove the pain itself [B.c]. We can even process and elaborate pain [D1], so that when we think about it, we will re-live it, avoiding potentially harmful situations.

1 At the end of the section, I present the translated version of a part of the schema proposed by Navarini (2020, p. 11), which outlines the various ways in which an individual can respond to and adapt to changing circumstances.

Anticipated Pain [A]	Anticipation or fear of pain
Accidental Pain [B]	Immediate pain reaction or pain Resistance [B.a] Removal of causes [B.b] Pain removal [B.c]
Processed Pain [D1]	Psychic trace of past pain [D1]

Tab. 2 Schematic summary of pain types (Navarini, 2020, p. 11, my translation)

Adopting behavioral and cognitive strategies to react to a certain experience is called "coping". Coping strategies are responses "aimed at diminishing the physical, emotional, and psychological burden that is linked to stressful life events and daily hassles. Therefore, by this definition, coping strategies are those responses that are effective in reducing an undesirable "load" (i.e. the psychosocial burden)" (Snyder, 1999, p. 5).

This does not imply that responding to pain or suffering is a simple or immediate process. As discussed in the previous chapter, receiving a cancer diagnosis typically initiates a complex process that unfolds through multiple stages and behavioral responses, including shock, anger, denial, bargaining, and other defense mechanisms.

Moreover, there is no guarantee that the strategies adopted will always be functional. Coping, in fact, may be either "adaptive" or "maladaptive"[2]. The term adaptive coping refers to behavioral patterns that allow the individual to face a specific problem, situation, or context in ways that are ultimately constructive. One example might be the recognition that the suffering one is experiencing is significant and requires support, prompting the individual to seek help from family, friends, or professional figures. Maladaptive coping, by contrast, includes responses such as ignoring or fleeing from situations that cause pain or distress.

Given that coping involves a process with distinct phases and is not necessarily positive or functional from the outset, psychological,

2 On adaptive and maladaptive coping see e.g. Endler & Parker, 1990; Folkman & Lazarus, 1980.

social, and spiritual support emerge as essential resources in this context. It is precisely for this reason that palliative care[3], especially in its early forms[4], may make the difference. Early palliative care, in contrast to commonly understood palliative care – which mainly addresses advanced or terminal in the final stages of illness – aims to accompany the patient from the time of diagnosis, improving both their quality of life and their overall well-being. To achieve this, palliative care adopts an integrated approach, addressing not only physiological but also moral, social, and spiritual dimensions. This ensures continuous and holistic attention to the individual's needs, including the support of both coping processes and the path toward acceptance.

The adoption of functional coping strategies is essential, as it is closely linked to positive adaptation. Numerous studies in the literature confirm this point. For instance, a research by Kathleen Dukes Holland and Carol K. Holahan shows that while positive coping strategies (e.g. positive reappraisal, planful problem-solving, seeking social support) are associated with beneficial adjustment, negative coping strategies (e.g. distancing, avoidance) correlate negatively with psychological well-being (Dukes Holland & Holahan, 2013). These findings underscore the importance of providing appropriate support to patients[5].

The communication of a diagnosis, particularly one concerning a chronic or degenerative illness, has a profound and transformative impact on a person's life. It inevitably reshapes their life and requires them to confront both actual and potential consequences.

---

3 For an historical review of this subject see Clark (2007). For an account on early palliative care see Tassinari et al., 2016.

4 The intuition that palliative care should not be confined to the terminal stage but may extended to other phases of illness is attributable to the surgeon Balfour Mount. Mount was also responsible for the introduction of the neologism "palliative care", which replaced the term "hospice" (see Navarini, 2020, p. 47).

5 In fact, while a positive reaction to illness can influence health outcomes (e.g. Stanton & Revenson, 2011), it has been demonstrated that "[n] egative emotions can intensify morbidity in chronic conditions such as cardiovascular disease, osteoporosis, and cancer" (Stanton & Revenson, 2011).

For this reason, while adaptation is certainly possible, it is a complex process that must be supported through the most appropriate tools and pathways. In the following section, I will examine how the experience of illness, in its transformative dimension, may serve as a catalyst for flourishing.

2. *Illness as a Transformative Experience*

As previously noted, the discovery of a chronic or degenerative condition entails profound changes: there is a clear "before" and "after" the diagnosis. It is precisely for this reason that illness has been described in the literature as a transformative experience. The philosopher Laurie A. Paul describes some types of experiences as "transformative" because they change the subject through the process itself (Paul, 2014).

Although Paul does not include illness among the experiences she defines as transformative, this connection has been advanced by Carel and Kidd in a recent work that aims at expanding Paul's concept. In particular, the authors argue that illness belongs to the category of "Nonvoluntary Transformative Experiences" (Nonvoluntary TEs), i.e. experiences that alter the subject even though the subject did not initiate or cause the transformative event. In addition to Nonvoluntary TEs, Carel and Kidd identify two other forms of transformative experience: "Voluntary TEs" and "Involuntary TEs". A voluntary TE may, for instance, be the decision to become a parent; an involuntary TE, by contrast, might involve losing one's sight while trying to protect another person. In such a case, although I did not intend to be injured, I did voluntarily place myself between the threat and the other person. Consequently, we may say I bear a kind of (non-strict) responsibility for what happened (Carel & Kidd, 2020).

As Carel notes: "Illness is a transformation that changes how and what one is and what one can aspire to, set as a goal, what horizons of living are possible within it. Serious illness has this transformative power. It is not one that leaves you intact. It is transformative, in the sense suggested by the philosopher Laurie Paul in her recent book Transformative Experience" (Carel, 2019, p. 156).

According to Paul's account transformative experiences can be so in two senses. First, they are "epistemically transformative" because they provide a degree of knowledge which was previously inaccessible. Illness is epistemically transformative in this sense: it is impossible to fully understand what it means to be ill without actually experiencing it. This is precisely why many doctors are deeply affected when they themselves become patients: there are no years of study or clinical practice that can teach what it feels like to receive a diagnosis, to undergo radiotherapy or chemotherapy, to face significant bodily changes, or to experience a recurrence of the disease[6].

Secondly, these experiences are also personally transformative, as they reshape the goals, desires, values, and preferences of the individual who undergoes them. In short, we could say that it is not possible to know in advance how a transformative experience will change us (nor to what extent) until we live through it. To define the sense in which a given experience may be personally transformative, Paul makes the following point:

> The sorts of experiences that can change who you are, in the sense of radically changing your point of view (rather than only slightly modifying your preferences), are experiences that are personally transformative. Such experiences may include experiencing a horrific physical attack, gaining a new sensory ability, having a traumatic accident, undergoing major surgery, winning an Olympic gold medal, participating in a revolution, having a religious conversion, having a child, experiencing the death of a parent, making a major scientific discovery, or experiencing the death of a child. The experience can be life-changing in that it changes what it is like for you to be you. That is, it can change your point of view, and by extension, your personal preferences, and perhaps even change the kind of person you are or

6 A particularly insightful contribution in this regard is offered by Robert Klitzman, who in his volume *When Doctors Become Patients* collects over seventy interviews with physicians who, having become patients themselves, reflect on the opacity of the illness experience for those who have not lived it firsthand. Klitzman's own reflections as a doctor-patient capture well the transformative nature of illness. As he writes: "The experience forced me to cross the border from provider to patient, and taught me how much I did not know" (Klitzman, 2008, p. 4).

> at least take yourself to be. If an experience changes you enough to substantially change your point of view, thus substantially revisioning your core preferences of revising how you experience being yourself, it is a personally transformative experience (Paul, 2014, p. 16).

What Paul outlines also applies to the experience of illness, during which projects, ambitions, dreams, and relational dynamics may undergo profound change. Even the body we once knew may no longer the same, and the world as we perceived may be irrevocably altered (Ricci, 2024b). In this sense, the experience of illness is also personally transformative.

Such transformability compels the subject to question their own narrative identity[7]. If illness indeed changes how we perceive reality, it may equally alter how we perceive ourselves.

These transformations are not necessarily negative. It is possible for patients to respond with feelings of loss, chronic sorrow, frustration, guilt, anger, loss of connectedness, struggling with change. Yet, there are also individual, who react positively to suffering, developing new character strengths and virtuous traits[8]. These constitute responses to life in illness that reflect positive adaptation. Lebacqz writes in regard to this: "The 'virtues' of the patient are qualities of excellence in response to the stresses of pain, discomfort, physical limitation, loss of autonomy, violation of privacy, vulnerability, and loss of self" (Lebacqz, 1989, p. 278).

This view is further supported by several studies, including that of Kyriaki Mystakidou and colleagues, which show that although illness can undermine one's life, relationships, roles, values, and beliefs,

---

7 When I use the term "narrative identity", I am referring to "the internalized narrative of oneself in relation to the world that provides someone's life with unity, purpose, and meaning" (Wright et al., 2020a, p. 62). On narrative identity see also at least McAdams, 2011; Ricoeur, 1991.

8 Carel holds that this also depends on the focus of the studies; in fact, "studies that focused on the experience of health within illness, led to radically different results. There seems to be a hidden dimension of health within illness that emerges when researchers ask patients explicitly whether they experience episodes of well-being or of feeling healthy within their illness" (Carel, 2019, p. 97-98).

confronting illness can also lead to positive change. Mystakidou and colleagues write:

> Many cancer survivors attribute positive outcomes or personal benefits to the cancer experience, including enhanced life appreciation, improved relationships with others, and positive changes in self-concept (Cordova et al., 2001). These findings in cancer populations join a larger literature suggesting that the effects of traumatic and stressful experiences are not uniformly negative and that various stressors including health problems (Katz et al., 2001) may be associated with profound positive consequences (Mystakidou et al., 2008).

Another particularly insightful study in this regard is the one conducted by Alastair Campbell and Theresa Swift in 2002. Their research was carried out in four different countries (Denmark, Great Britain, Italy, and the Netherlands) and involved 44 patients affected by various chronic conditions, including rheumatoid arthritis, endometriosis, end-stage renal disease, and mood disorders. The results clearly show that positive adaptation to illness, accompanied by the development of virtuous traits, is indeed possible[9].

This highlights at least two key aspects: (i) virtuous traits may emerge as a consequence of a positive reaction to illness, and (ii) a well-developed or flourishing character can support patients in coping with suffering, and with the transformations and changes brought about by chronic or degenerative conditions. Campbell and Swift note: "The themes reflect not only virtues that have developed as a consequence of the onset of illness, but also existing virtues that are drawn upon in order to face the challenge of illness" (Campbell & Swift, 2002, p. 31).

To briefly summarize what has been discussed so far in this chapter, illness constitutes a traumatic experience that clearly demarcates a "before" and an "after." It enters the person's life as an intruder, forcing a revision of one's life trajectory and often also of one's personal identity. In the process of confronting these changes and

9 The subjects of Campbell and Swift's study have indeed shown traits such as courage, realism, self-respect, sense of humor, hope, and the ability to maintain good relationships with others.

the consequences brought about by illness, the patient begins a path of coping and adaptation, which unfolds through several stages. For this reason, support from both the relational network and healthcare professionals is essential.

While it is certainly possible for a person to respond negatively through maladaptive coping strategies, experience also shows the potential for a positive response to illness. As a result of this kind of positive adaptation, the literature indicates that one of the possible outcomes is a change in character and the development of new traits. This form of adaptive coping, which may lead to the emergence of new character strengths, is referred to in the literature as posttraumatic growth.

The term posttraumatic growth (from now on PTG) was coined by Richard Tedeschi and Lawrence Calhoun (e.g. 1996) and it stands for a positive psychological change that can follow from having struggled with very challenging life circumstances (Calhoun & Tedeschi, 2006; Tedeschi & Calhoun, 2004). As Aliza Panjwani and Tracey Revenson notes, PTG generally brings with it many positive changes: "[p]eople who report PTG find themselves transformed as a result of illness, often by building a reservoir of personal strength, acquiring a better appreciation of life, or strengthening relationships with family and friends" (Panjwani & Revenson, 2021, p. 66).

According to the literature, PTG can occur in at least one of these dimensions: (i) *social*, e.g. increased compassion for others, improvement in social resources; (ii) *cognitive*, e.g. increased self-conception, perception of personal strengths, improvement of problem-solving skills; (iii) *emotional*, e.g. feeling more relaxed; (iv) *physical*, e.g. changing lifestyle; and (v) *spiritual/philosophical*, e.g. increased appreciation of life[10].

It is important to remember that it is not suffering itself that brings about positive changes; otherwise, anyone suffering would be transformed and strengthened by it. Conversely, it is the subject's *response* to that suffering which determines whether they will get anything good out of it (Wright et al., 2020a)[11]. Consequently, the

10 For a review, see Mystakidou et al., 2008.

11 It has also been shown recently that posttraumatic growth appears to be promoted by *reacting copying*. In a recent paper by Fujimoto and

growth and development of new traits and dispositions should not be taken for granted. It should be thought of as the result of the subject's work on their inner world.

Let us consider a concrete example and imagine a subject who receives the news that they have cancer. It is reasonable to assume that they will go through various stages of emotional distress, experiencing negative emotions and suffering along the way. At first, they may cope with this new experience by denying what is happening to them, until they come to acknowledge the reality of their condition. When this happens, as human beings naturally tend to life and not to death (Navarini, 2020, p. 25), may feel as though they have fallen into a "trap" from which there is no escape. The inability to flee from a situation perceived as threatening may trigger intense emotional distress[12]. However, if the subject is supported through the various coping stages, they may gradually come to reinterpret what initially seemed like a threat, not as a trap, but as a challenge. By accepting their new condition and adapting positively to it, they may experience post-traumatic growth (PTG), ultimately developing new character strengths and moral dispositions.

Okamura, the authors resume the fundamental differences between conventional and proactive coping as follows: "Regarding the differences between conventional and proactive coping, Usami (2012) pointed out the following three points: (i) while conventional coping is an effort to deal with stressors that have already occurred, proactive coping is directed to the future and includes efforts to promote challenges and personal growth; (ii) while traditional coping mainly involves risk management when a negative appraisal of threats and harms is made on stress, proactive coping involves goal management with stress as an opportunity for challenges and growth and (iii) while conventional coping is triggered by negative appraisals on requests from the environment, proactive coping is triggered by more positive motives. However, research on proactive coping is currently scarce (Sleczka et al., 2016; Bhattacharyya et al., 2018; Rai et al., 2019; Russo et al., 2019) and to our knowledge, there is no research available on proactive coping in subjects with cancer (Lisica et al., 2019)" (Fujimoto & Okamura, 2021, p. 86).

12 This is particularly evident in the case of chronic/degenerative diseases and even more in the advanced and terminal stages of the latter, where symptoms become more present, and pain and suffering may become a dimension to be confronted on a daily basis.

In this sense we can say that life offers us circumstances for growth, but it is up to us to respond to them in a way that allows flourishing. Yet such a process is never immediate, nor effortless. Change is often destabilizing, and the path it requires can be long and demanding.

One of the factors that makes coping with serious illness particularly challenging is the series of bodily changes that often accompany it. Illness affects not only the body but also the person's lifestyle, emotional state, and perception of self. Consider, for example, patients with breast cancer. They may face the amputation of a deeply intimate part of their body and witness it undergo visible and symbolic transformation. Chemotherapy may lead to the loss of hair, eyelashes, and eyebrows, while radiation therapy may result in permanent changes to the skin. These visible transformations are compounded by less apparent but equally impactful changes, such as chronic fatigue or a general loss of energy. Such shifts are rarely easy to accept. On the contrary, they can deeply reshape a person's life and, in many cases, their very sense of identity (Ricci, 2024b).

These aspects can make coping and adaptation particularly complex. Bodily transformation affects both future expectations and self-image, requiring the individual to engage in a process of self-acceptance and redefinition of their narrative identity. As we have seen, changes in the lived body operate on a deep existential level. Consider, for example, the use of assistive devices like a wheelchair or a stick. These are not merely tools; they change not only the subject's physical relation to the world, but also how they are perceived and how they perceive themselves.

To illustrate this more clearly, take the seemingly simple case of someone who becomes nearsighted and starts wearing glasses. For a person who has always had perfect vision, eyeglasses may initially feel foreign. In the beginning, they may not even recognize themselves in the mirror. Over time, however, those glasses may become so integrated into their identity that they cannot imagine themselves without them. They may even forget they are wearing them or fail to recognize themselves when changing frames. What has changed? Their self-perception and self-recognition.

This change is not automatic, it requires adjusting to vision loss, coping with the associated feelings, and learning to accept a new image of oneself. It also demands reappropriating a new narrative identity.

Merleau-Ponty offers valuable insight into the relationship between bodily change and identity: "The blind man's stick has ceased to be an object for him and is no longer perceived for itself; its point has become an area of sensitivity, extending the scope and active radius of touch, and providing a parallel to sight" (Merleau-Ponty, 1962, p. 143).

This reflection reminds us that when our body changes, so does our way of inhabiting the world and understanding ourselves. Often, we think of illness as producing limitations alone. Yet the body is also a medium of self-recognition and meaning. Bodily changes, therefore, may profoundly affect the person's sense of self and generate suffering. For this reason, these experiences must be acknowledged, and the person supported through them.

To understand just how deeply illness-related bodily transformation can affect a person, let us consider the case of a young athlete, Harry, who has always envisioned himself as a future professional runner. Upon receiving a diagnosis of facioscapulohumeral muscular dystrophy (FSHD), a progressive condition typically appearing in adolescence, Harry will inevitably need to revise his dreams, expectations, and personal goals. The progressive loss of motor function associated with FSHD may lead him to experience his body as unfamiliar, even alien, and result in a sense of narrative disintegration (Cassell, 1991)[13].

This example explain why it is vital to address the ill person's needs on psychological, moral, social, and spiritual levels. Bodily changes can be a source of deep suffering, but a holistic and integrated form of care can help individuals cope more effectively and preserve a sense of coherence and purpose. Taking Harry's case as an example, we can imagine that if he were accompanied from the outset by a team focused on his needs from multiple perspectives, the process of

13 Bury (1989) also described illness as the disruption of someone's biographical identity.

adaptation, though difficult, would be more sustainable. Over time, he may succeed in reconstructing the sense of self that the diagnosis initially disrupted and go on to live a good, full, and meaningful life.

3. *PTG, Virtues, and Personal Identity*

Literature has, in fact, shown that a positive adaptation to illness is possible, and can involve meaningful transformations, such as the strengthening of character and the development of positive character traits. These traits may foster the subject's well-being, as they may allow individuals to live better lives in illness. For example, perseverance is a crucial trait, as it can help patients rediscover the motivation needed to pursue daily goals (such as adhering to treatments or engaging in regular physical activity) despite the inherent difficulties (Kassim et al., 2021). Therefore, although it would be unrealistic to expect that every individual responds to severe trauma by developing new virtues, this outcome remains a concrete and attainable possibility. Jennifer Cole Wright, Nancy Snow and Michael T. Warren write about this: "suffering provides an opportunity to develop [virtues] – an opportunity that people may not get elsewhere" (Wright et al., 2020a, p. 65).

In *Suffering, Virtue, and Character: Why the Science of Virtue Matters*, they emphasize that although experiences of deep suffering can be traumatic, they may also facilitate the development of certain traits by altering our "narrative structure". In fact, according to them, suffering represents a distinct opportunity for the development of new positive dispositions, insofar as it confronts the subject with virtue-relevant stimuli, such as the need to be brave or to persevere. To cite an example offered by the authors, dealing with illness requires one to develop greater problem-solving skills, the ability to observe a situation from different angles, and the capacity to reflect carefully. Repeated engagement in such contexts may gradually foster the development of practical wisdom. Furthermore, illness does not confront the patient with such situations only once, but repeatedly. Thus, the more the subject is exposed to contexts in which practical wisdom is required, the more opportunities they have to exercise and refine it. In this sense, situations that require the activation of *phronesis* can gradually reshape the subject's personal and narrative identity.

It is crucial to emphasize again that for such transformation to occur, positive adaptation to the experience of illness is essential. This highlights once again the importance of the pathway in which the patient is embedded and the quality of the care they receive. While repeated exposure to morally significant challenges may support the development of certain traits, it also requires a flourished character, capable of recognizing the appropriate response. If this condition is met, experiences of suffering may, over time, contribute to character development. Nonetheless, this presupposes the patient internal openness and readiness. As Navarini argues, there are two fundamental ways in which an individual can approach the experience of illness: it "can be simply 'suffered' [...], or it can be invested, by rewriting the subject's priorities and disposing them to a constant creative work to shape themself and their relationships. The capacities that emerge from this practice of suffering are intrinsically moral" (Navarini, 2020, p. 171, my translation). In this sense, it should be clear that illness does not hinder moral qualities; rather, it exposes the subject to a set of circumstances that can even "foster" their cultivation.

This does not make the acquisition of the virtues easy or immediate. On the contrary, as I have emphasized, it requires adequate support during difficult moments and a comprehensive care approach that addresses all dimensions of the person's needs. This also entails the possibility of engaging with the patient's inner world, not only by listening to their needs and understanding their emotional responses, but also by offering structured pathways for character work. Character education programs could provide a structured, continuous practice focused on specific traits, which is essential for flourishing. Without deliberate practice, no one becomes virtuous.

4. *Suffering as a Virtuosity Activator*

To delve deeper into the connection between illness and flourishing, the idea that virtues can arise from effective coping and positive adaptation is supported by various models, both psychological and philosophical. One of the most compelling perspectives that bridges these two approaches is that of Wright, Snow and Warren. The authors take a strong stance by arguing that not only is it possible to flourish while suffering from illness, but that suffering itself can serve as a powerful catalyst for virtues and may even alter an individual's personality structure.

The starting point of this account is the "Whole Trait Theory" (WTT) developed by Will Fleeson and Eranda Jayawickreme (e.g. Fleeson & Gallagher, 2009; Fleeson & Jayawickreme, 2015; Jayawickreme & Blackie, 2016), which Wright, Snow and Warren aim to integrate with Aristotelian virtue theory (Wright et al., 2021, p. 6). As the authors explain, according to WTT, personality traits can be described as:

> [C]omposed of a set of situation-specific trait-appropriate responses (typically a combination of mental states and behaviours), which are produced when certain social-cognitive mechanisms (i.e. cognitive, affective, motivational states, processes, capacities, and structures) are triggered by the perception of trait-relevant stimuli. The degree to which a person can be said to possess a particular personality trait is determined by the "density" of her trait-appropriate responses – in other words, by the frequency (and consistency) with which she produces trait-appropriate responses when encountering trait-relevant situations and by the range of different types of trait-relevant situations in which she produces them (Wright et al., 2019, p. 59).

Thus, to determine whether a person possesses a particular trait (e.g. friendliness), WTT requires measuring how frequently and consistently they respond appropriately across a wide variety of situations that should activate that trait. This entails that the more a person exhibits friendly behavior in relevant contexts, the more friendly they are.

Wright and colleagues propose that virtues function in a similar way. The more a person responds to trait-relevant situations with virtuous actions, the more likely it is that they will internalize that trait, stabilizing it over time. The authors write on this: "[A] person's virtue (let's say, her generosity) can be measured as the distribution of her "generous" responses (i.e. thoughts, desires, emotions, motivations, and actions that are generosity-appropriate) to a variety of generosity-relevant stimuli presented across a wide range of situations over time" (Wright et al., 2019, pp. 59-60).

Returning to the relationship between suffering and virtue, the authors observe that certain circumstances, such as those involving adversity or illness, are particularly conducive to cultivating virtuous traits because they repeatedly activate them. To illustrate this, they discuss the example of a person who, through suffering, comes to understand the nature of a loving relationship. Because they have both provided and received support during hardship, they may develop a schema of what a loving relationship should be. Furthermore, because they have experienced the importance of those and their positive impact on themselves and others, they may wish to see these behaviors more often in the world. Over time, these behaviors become internalized and ultimately reshape the individual's personality (Wright et al., 2019).

In other words, we may say that exposure to demanding circumstances can lay the groundwork for lasting moral transformation. Once integrated, these changes do not merely constitute temporary adjustments; they become enduring components of one's character and a resource for flourishing.

According to Wright and colleagues, virtue acquisition occurs through three steps: (i) the *perception of trait-relevant stimuli*; (ii) the *activation of underlying socio-cognitive mechanisms* (such as beliefs, schemas, desires, and values); and (iii) the *generation of appropriate responses* (Wright et al., 2019, p. 56). This process is particularly likely to be activated in challenging or traumatic experiences, as these significantly increase the frequency and salience of stimuli relevant to specific traits. The more a subject is exposed to these stimuli, the more likely they are to activate the corresponding cognitive-affective responses, leading to the reinforcement and eventual stabilization of the trait. Repeated exposure and appropriate response patterns may result in the individual becoming increasingly expert in the activation of that trait, even in complex situations[14]. In this sense, flourishing is

14 Wright and colleagues specify this further by writing: "As we become more attuned to virtue-relevant stimuli, we become more aware of the full range of stimuli that are present, as well as the many ways this range of

defined by Wright, Snow, and Warren as "an increase in the person's sensitivity to trait-relevant stimuli" (Wright et al., 2020, p. 63).

To clarify their view, the authors present the case of students exposed to severe trauma (e.g. civil wars) over the previous decade. The exposure to such events may have led these subjects to develop a certain degree of sensitivity to certain virtue-relevant stimuli. This would also increase their awareness and generate an amplified attunement towards them. Without that experience, they might never have become so sensitive to these needs.

Wright and colleagues' account illustrates that flourishing is possible even in the experience of illness, as particularly difficult circumstances can serve as rich stimuli that activate specific traits. A person frequently exposed to situations that require perseverance, for instance, has numerous opportunities to refine and consolidate that trait.

This view provides a valuable counterargument to those who claim that requiring ill individuals to develop virtues places an undue burden on them (see Kidd, 2012, p. 513). Instead, it suggests that suffering can create the very conditions that make virtue development possible, provided that the subject is supported in responding constructively. Still, it is vital to reiterate that this process is not automatic. Positive adaptation depends on many factors, including the care pathway the patient is involved in and the quality of the support they receive. For this reason, early palliative care can be essential, not only for medical and psychological benefits, but also for enabling patients to reinterpret their condition, access new tools for navigating their experience, and potentially grow through it. This flourishing, which transcends the limits that illness and suffering bring, is fundamental in so far as it enables the ill subject to better manage life in illness.

If this is correct, then illness and suffering are not simply obstacles to human development. Rather, under the right conditions, they may also conceal a positive potential for the subject.

---

stimuli interact-complementing and contradicting one another-blending together into a complex and nuanced call for virtuous responding. And in learning to navigate this rich environment of virtue-relevant stimuli, we develop the capacity to regulate our responses so as to be more appropriate for this higher level of complexity" (Wright et al., 2020a, p. 58).

5. *Accounts on Suffering and Flourishing*

Wright, Snow and Warren's account on suffering is not the only approach that assigns a positive meaning to this experience. In the literature, several authors have highlighted the possible connection between suffering and flourishing.

These approaches can be categorized into two groups: "instrumentalist accounts" and "edificationist accounts". Instrumentalist views tend to regard suffering as a necessary means to flourishing. A paradigmatic example is offered by Nietzsche, who argues that suffering is not only fundamental for the strengthening of character and the development of virtues, but is also essential for achieving excellence of character. From this perspective, instrumentalists place a strongly positive value on the experience of suffering in all its forms, treating it as an indispensable resource in the process of human flourishing.

Edificationist accounts emphasize instead the positive potential of circumstances such as illness, without considering them essential to flourishing. Edificationists recognize that certain experiences of suffering may amount to genuine trauma and, in some cases, may be too brutal for the subject to derive anything good from them. Nonetheless, they argue that even the most traumatic conditions can conceal a positive potential. Illness and suffering, from this point of view, are not to be considered as necessary instruments of character development, but rather as contexts in which flourishing remains a concrete possibility.

In the following paragraphs, I will analyze in detail some of the most representative accounts of both the instrumentalist and edificationist perspectives.

1.5. *Instrumentalist Approaches*

One of the most significant contributions to the instrumentalist view is that of Friedrich Nietzsche[15]. Nietzsche maintains that

15 Although Nietzsche's account is very much focused on the phenomenon of suffering and its usefulness to human beings, this is not the only example. The Cynics, such as, for example, Diogenes Laertius (Diogenes

experiences of suffering enable personal growth and represent a privileged pathway to human flourishing. For this reason, suffering should not be avoided, but rather welcomed and deliberately sought as a means of personal growth as strength is given as the result of positive responses to it[16]. According to some interpretations, Nietzsche seems to conceive of suffering as the only means by which an individual may become virtuous[17].

What Nietzsche wants to highlight is the inevitability of suffering as a condition of human existence. More than that, the German philosopher holds that suffering should sometimes even be pursued, because it is only through suffering that one can attain excellence. For Nietzsche, excellence that does not pass through suffering is inconceivable.

Another philosopher who develops a similar perspective is Brady. Drawing on two key aspects of Nietzsche's view, namely, (i) the idea that psychological strengths depend on a positive attitude, and (ii) that suffering is essential to life and to the realization of its value (see Brady, 2018, p. 94), Brady arrives at comparable conclusions. He argues not only that suffering can possess both intrinsic and instrumental value, but also that, when experienced in the right degree and under the right circumstances it can itself be "virtuous"[18]. Brady, in fact, emphasizes that suffering must be faced and accepted, and that it is impossible to become virtuous and flourish without undergoing it[19]. In his words: "I'll outline the truth in the idea that

---

Laertius 6.104 and 7.122), considered it as essential for moral flourishing, for the development of virtues, and ultimately for "asceticism".

16 For more on this topic, see Brady, 2018.

17 Due to his distinctive view of the relationship between suffering and flourishing, Nietzsche has been regarded by some scholars as a virtue ethicist, despite the fact that he never developed a systematic account of virtue ethics (e.g. Solomon, 1999; Swanton, 2015).

18 Brady writes about this: "On this account, a disposition to feel pain (to the right degree, in the right circumstances) in response to physical threat or injury constitutes an excellence of the pain system, a feature of the system that is vitally important in enabling it to accomplish its valuable goals: the avoidance of threat and the repair of damage" (Brady, 2018, p. 86).

19 Brady argues that illness is fundamental to understanding the fragility of human life. He writes is this regard: "it is only when our pride is hurt by

pursuing suffering and adversity can contribute to one's virtue and happiness, since suffering and adversity can be a condition of certain goals and ends" (Brady, 2018, pp. 94-95).

According to Brady, suffering is also a necessary condition for the development of various virtues, including what he calls "the virtues of strength and vulnerability", as well as moral virtues (e.g. compassion) and social virtues like justice, love, and trust. While he acknowledges that suffering and illness can be deeply distressing and may, in some cases, contribute to a weakening of character, Brady insists that they can also serve as powerful resources for deepening and refining the individual's inner world. He further notes that illness has the potential to make individuals less conceited and thus able to understand how fragile our and others' lives are (see Brady, 2018, p. 154).

5.2. *Edificationist Approaches*

5.2.1 *Havi Carel*

Having now examined the instrumentalist accounts; I will turn to a more in-depth analysis of the edificationist perspectives.

An influential edificationist account is offered by Havi Carel, who has extensively explored the role of suffering and illness in relation to health, moral flourishing, and happiness. Carel begins from the premise that illness may represent a real trauma, and, like all traumatic experiences, it requires a revision of the ill person's identity as well as adjustments. In this sense, according to the philosopher (who echoes Paul's account), illness is a life-transforming process that "induces adaptation" (Carel, 2019, p. 19).

Although coping with this experience is difficult because it implies confronting and accepting significant worsening changes that introduce a greater level of complexity into the ill person's life, Carel shows that illness does not exclude the possibility of

---

illness, when our lack of ability is a source of despair, that we may be inclined to adopt a less vain and self-aggrandizing posture, realizing the fragility of our life and our good fortune at this point" (Brady, 2018, p. 109).

experiencing elevated levels of well-being[20], living a good life, and even flourishing. Illness can force us to rethink our idea of "health", our concept of the good life, inviting us to reevaluate these dimensions and to reconsider ourselves as fragile. In this sense, Carel emphasizes that illness does not represent a threat to one's capacity for a meaningful life.

Those unfamiliar with the experience of illness may assume that well-being is unattainable without certain prerequisites, such as perfect health, full autonomy, or absolute freedom, leading them to believe that ill individuals are necessarily excluded from flourishing. In contrast, reality suggests otherwise: many patients report levels of well-being and perceived quality of life that far exceed what one might expect from the outside (Carel, 2019, p. 96). Particularly significant in this respect is a study by Alexa Stuifbergen and colleagues, which aimed to assess the level of health perceived by individuals living with disabilities (Stuifbergen et al., 1990). The findings are striking: 73% of the respondents rated their health as good or excellent, revealing how the concept of health can vary considerably from person to person. For some, being healthy meant not needing medication; for others, it meant being able to care for themselves independently.

In articulating her view on illness and flourishing, Carel gives special prominence to two traits: *adaptability* and *creativity*[21]. Carel describes adaptability (which she distinguishes from adaptation, as this is a term generally intended to indicate an evolutionary process)[22]

---

20 Throughout her work, Carel tends to employ the term health rather than well-being. The notion of health within illness was primarily developed in the nursing literature, particularly in the writings of Lindsey and Moch (e.g. Lindsey, 1996; Moch, 1989), who argued for the positive transformative potential of illness. According to Carel, these perspectives demonstrate the possibility of experiencing a form of health even while being ill (Carel, 2007, p. 103).

21 Carel does not explicitly employ the concept of virtue in her work. However, Kidd interprets the two traits that Carel identifies as central to a positive response and adaptation to illness, namely, adaptability and creativity, as bearing a clear virtuous connotation.

22 As Carel explains, adaptation "denotes an alteration in the structure or function of an organism or any of its parts that results from natural selection" (Carel, 2007, p. 105).

as "the behavioural flexibility enabling ill or disabled people to adjust[23] their behaviour in response to their condition" (Carel, 2019, p. 100). Creativity, on the other hand, consists in the ability to find a new way of responding to circumstances, despite impediments that make the "normal" administration of those circumstances more complex or even impossible. Adaptability and creativity are therefore essential resources for the ill person, as they may enable them to adjust in a meaningful way and improve their quality of life,

According to Carel, adaptability can manifest physically but also extends to other domains of life, including daily routines, social interactions, and time management. For instance, the ill person must adjust to new rhythms as well as to a transformed daily life and altered ways of relating to the world and to others. These changes require the cultivation of various nuances of the same overarching trait of adaptability. The development of adaptability and creativity not only facilitates adjustment to the new condition but also contributes to a positive re-signification of illness, suffering, and limitation, and enables the subject to adopt a constructive attitude toward the future.

Finally, according to this account, not only are well-being and happiness compatible with illness, but, as the philosopher explains in her paper *Can I be Ill and Happy?*, the flourishing of character can actually help the subject to cope better with their condition. Carel writes in this regard:

> Illness is a deep change, and as such can teach humility, forgiveness (often to ourselves), patience, acceptance, and fortitude, amongst other virtues. Courageous acceptance, resilience and reflective coping may enable the ill person to bear their illness well. However, as Ian James Kidd (2012) argues, that does not mean we should seek out illness; it is a misfortune. But although illness is unwanted and unwelcome, one's reaction to such adversity may cultivate excellence by eliciting particular responses to it (Carel, 2016, p. 243).

---

23 Adaptability, according to Carel, would be more or less "spontaneous" depending on circumstances and conditions. To explain what she means by this, the philosopher brings up the example of "panting", which is a natural and spontaneous response in terms of the body's adaptability to oxygen deprivation (see Carel, 2019, p. 101).

5.2.2 *Ian J. Kidd*

Another edificationist approach is that of Ian Kidd. Like Carel, Kidd begins by recognizing illness as a traumatic event that can bring about a series of dramatic changes and transformations. Precisely for this reason, the philosopher emphasizes that illness and suffering carry a certain value. In particular, they can be a source of *existential value*, teaching the patient to better understand both their limitations and their potential. They can also offer *cognitive value*, providing artistic and philosophical inspiration or encouraging the development of positive dispositions such as empathy towards other ill individuals (see Kidd, 2012, p. 509).

According to Kidd, illness is also significant because it may foster "distinctive opportunities" for the cultivation and development of virtues that help the patient better cope with their condition, both practically and psychologically. Kidd writes in this regard:

> Consider a series of typical virtues. An *adaptable* and *creative* person can better respond to their illness than an inflexible, increative person, especially if they are also *patient* and have *fortitude*. Conforming to drug regimens and exercise requires *discipline*, while engaging with healthcare professionals invites *trust* and *honesty*, as well as *hope* concerning possibilities for future health and *gratitude* for what health one does have. (Kidd, 2012, p. 502)

It is worth noting that according to the philosopher not every illness will necessarily lead to flourishing. Kidd argues that its edifying potential depends on three key factors: (i) *the nature of the illness*; (ii) *the prior ethical character of the person*; and (iii) *their social environment* (Kidd, 2012, p. 504).

Regarding the first criterion, the potential for positive transformation depends on the intensity, duration, and cognitive impact of the illness. Kidd notes that different types of illness will produce different transformative effects. For instance, a mild fever is far less impactful than a serious illness like cancer. In some cases (such as with Alzheimer's disease), the cognitive impairments may be so severe that flourishing becomes impossible.

The second criterion is the prior ethical character of the person. Kidd argues that a person's temperament and virtues can significantly influence how they respond to life challenges, including illness. A pessimistic disposition, for example, may hinder positive transformation. This view is shared by Ruini and Vescovelli (2013), who maintain that virtuous individuals are more likely to find meaning even in painful and traumatic experiences, thus fostering PTG and better coping. However, this should not be taken for granted. Even a virtuous person may be so traumatized by illness that they "go to pieces", while a previously vicious individual may experience a profound moral transformation (Kidd, 2012, p. 505).

The third factor is the social environment. According to Kidd, supportive relationships can significantly influence a patient's ability to flourish by facilitating the development of virtues. Conversely, a toxic or neglectful environment can undermine confidence and impede growth. Therefore, a supporting social environment can, according to the author, radically change the way in which the patient will react to their condition[24].

Kidd acknowledges that the edificationist view may be subject to several objections. He therefore addresses a number of critical concerns, namely, (a) the instrumentalist objection and (b) a set of objections including the novel demand, failure, and Medusa objections (Kidd, 2012, pp. 206-216).

A first objection that could be made from an instrumentalist point of view is that (a) illness is actually a good or even desirable condition and that, as such, it provides the subject with unique chances for moral development and for the cultivation of virtues. This could be gathered from strong instrumentalist accounts like Nietzsche's or the Cynics', who believe that "certain forms of human good are only available through experiences of illness" (Kidd, 2012, p. 507). Kidd argues that this objection can be countered by highlighting the numerous opportunities for flourishing that exist beyond illness.

24 Kidd takes this point from the third chapter of Carel's (2019) book in which the philosopher talks about the "social world of illness", emphasizing how much the ill person's loved ones have a power over the way they will respond to illness.

Consequently, being ill is not an essential prerequisite for a person's flourishing, as even in a so-called "healthy" life, there are many occasions that can foster personal growth and development (Kidd, 2012).

Kidd does not deny the transformative potential of illness, but he emphasizes that illness is not the only experience capable of fostering moral growth. Moreover, he critiques the idea that suffering should actively seek to build character. Conversely, he describes instrumentalist views as "hyperbolic, ethically misguided, and imprudent" (Kidd, 2012, p. 512).

The second set of possible objections (b) focuses on the idea that the edificationist's proposal should not be promoted, but rather discouraged first because since (i) illness is intolerable, asking patients to work on their character adds undue stress. On the contrary, (ii) it should be avoided exposing patients to the risk of failure (which is a real possibility in the case of flourishing), as there is no guarantee that attempts at flourishing will succeed. Third (iii), encouraging virtue development could lead to implicit moral judgments against those who do not succeed, potentially leading to unjust healthcare practices.

Kidd responds to the three objections (Kidd, 2012) by noting flourishing is not an obligation but a possibility; character development is difficult even for healthy individuals and its success depends not only on individual effort but also on external factors such as the patient's character, the nature of the illness, as well as the social environment. Therefore, since success (which never depends solely on the subject) is never guaranteed under any circumstances, what really matters is not becoming virtuous but to *try* to flourish. Moreover, if flourishing does not depend entirely on the subject, why should the subject feel like a failure, be morally condemned, or condemn themself if they fail? As Kidd concludes "[T]he edificationist account of illness, at least as I have developed it here, should not include the strong normative claim that an ill person is, by the fact of their illness, thereby placed under an obligation to exploit the edifying potential of their illness" (Kidd, 2012, p. 516). Flourishing thus remains, according to the philosopher, a concrete option for the subject and not a cruel demand or an unbearable burden.

5.2.3 *Jennifer C. Wright, Nancy Snow, and Micheal T. Warren*

Another noteworthy contribution that can be classified within the edificationist accounts is that of Wright, Snow, and Warren. As already discussed, Wright and colleagues observe that, although experiences of suffering often constitute traumatic events, they do not preclude the possibility of moral growth. Conversely, the latter can arise from post-traumatic growth (PTG). In this view, suffering appears as a potential catalyst for the transformation of an individual's personality traits[25].

According to the authors, this process may occur because experiences of suffering increase the number of stimuli that can activate certain traits. Exposure to it (such as in the case of illness) may heighten the subject's sensitivity to trait-relevant stimuli, prompting them to respond with the appropriate character trait more consistently. Over time, repeated virtuous responses to similar stimuli may reinforce the trait, facilitating its integration into the individual's personality.

To better understand this model, consider the case of Hannah, a middle-aged woman who has been diagnosed with Parkinson's disease. Her daily therapy involves a complex and demanding routine: she must take multiple medications at precise times, engage in physical activity, and attend physiotherapy sessions. Each of these activities, although crucial to her well-being, requires significant effort and dedication. Taking medication regularly demands discipline and attention; exercising every day, despite physical discomfort and fatigue, requires resilience; and committing to physiotherapy calls for perseverance, especially when improvements are slow and difficult to perceive.

---

25 This idea that suffering can be a path to virtues is also shared, as we will see later, by Lebacqz, who writes about prudence: "Ironically, illness and the accompanying role of patient can themselves be the path to prudence" (Lebacqz, 1989, p. 283).

Hannah's condition is particularly limiting: her motor functions are impaired, and simple movements often require great effort. Sometimes, to perform a single exercise correctly, she must repeat it several times, battling against her body's rigidity and unresponsiveness. This struggle can be discouraging, and moments of frustration are frequent. Yet, she continues with her regimen, gradually cultivating the ability to endure, to stay focused on long-term goals, and to maintain hope even when results are not immediate.

This example clearly illustrates that, compared to a healthy individual, Hannah is exposed more frequently to circumstances that require perseverance and consistency. According to Wright and colleagues, this repeated exposure trains the subject to recognize and respond to trait-relevant stimuli. Over time, this may reinforce t the subject's character. In Hannah case, perseverance and discipline may become deeply embedded traits, transforming not only her response to illness but also her broader outlook on life.

A pivotal virtue in this model is practical wisdom (*phronesis*), which the authors regard as crucial in helping individuals discern the most appropriate behavior in varying contexts, and to adapt and respond effectively, even in highly complex or emotionally charged situations. Suffering, in particular, is identified as a powerful activator of this virtue as: "[g]rappling with difficult situations, figuring out how to overcome obstacles and survive loss requires problem-solving, perspective-taking, careful reflection and analysis, and developing an honest accounting of our lives call for the exercise of practical wisdom" (Wright et al., 2020, p. 65).

These considerations position Wright and colleagues firmly within the edificationist tradition. While they do acknowledge that experiences of suffering may be conducive to flourishing, they do not regard it as a necessary precondition for the development of character. On the contrary, they explicitly recognize illness as a traumatic experience. However, despite its traumatic nature, Wright, Snow, and Warren argue that illness may conceal otherwise unexplored opportunities for flourishing.

5.2.4 *Claudia Navarini*

A final account that deserves careful consideration within the edificationist framework is that of Navarini. In the introduction to her *Cure palliative Simultanee e sviluppo delle virtù* [Early Palliative Care and Virtue development, my translation], Navarini questions whether and how it is possible to speak of moral development when life is coming to an end and "when medicine "withdraws" into palliation" (Navarini, 2020, p. 5, my translation). The aim of the philosopher's text is in fact to investigate whether the questions raised by the pain and suffering that accompany chronic degenerative illnesses can be addressed not only by palliative care, particularly in its early form, but also through the cultivation of a virtuous character.

While starting from the premise that illness represents a traumatic experience that brings significant changes in the subject's life on a physical, moral, and psychic level, Navarini argues that it can also represent a space for moral growth and a place for the cultivation of the virtues. In this sense, her work[26] – together with some subsequent contributions (e.g. Navarini, 2022) – belongs to the edificationist tradition. Although she recognizes that some life conditions may be

26 In this regard, Navarini integrates the virtues identified by Carel and Kidd, presenting them as dimensions of practical wisdom (Navarini, 2020, p. 136). As she explains, "I consider the trait of adaptability in particular as a top-down feature of practical wisdom, that is, a particularizing dimension in which ethical principles are 'adapted' to the situation. This characteristic of practical wisdom – its capacity to deduce and apply general principles to particular cases – is quite evident in simple situations, aligning with the Aristotelian idea that phronesis consists in grasping the good in concrete terms. However, in more complex and delicate decisions, the wise person does not merely apply the principle deductively, but adapts it in such a way that, in order to fully align with the concrete situation, they may refine or even reinterpret the moral principle – at times appearing to violate the rule, while actually fulfilling it at a higher level. In such cases, one may speak of a top-down, adaptive, rather than deductive, function of practical wisdom" (Navarini, 2020, pp. 136-137, my translation).

pervasive, long-lasting, and obstruct flourishing, this does not negate the possibility for character development.

Navarini supports this view with reference to empirical studies (e.g. Campbell & Swift, 2002), which show that virtues can emerge from functional adaptation to illness and may significantly improve a patient's ability to live well despite their condition. Accordingly, even life in illness can be good.

Central to Navarini's proposal is a triad of virtues: hope, courage, and *phronesis*. Among these, hope plays a particularly critical role. It allows the subject to reorient themselves when original goals become unattainable, directing them toward alternative paths that are neither illusorily easy nor unrealistically out of reach, thus avoiding false hopes. As Navarini notes, hope is often future-oriented and might therefore appear ill-suited to the experience of serious illness. Nevertheless, she writes: "[T]he experience of ill persons shows that their sensitivity to hope is remarkable, and that it is in fact one of the moral traits they most feel the need for, in order to find meaning in a life marked by suffering" (Navarini, 2020, p. 129, my translation).

Importantly, hope in this context is not confined to the prospect of healing. It may concern residual well-being, care for loved ones, or the pursuit of short-term goals deemed meaningful, such as seeing a relative, attending a significant event, or telling one's life story. Hope helps the ill person to perceive their life, despite its fragility and precarity, as still rich in meaning. Unsurprisingly, Navarini notes a well-documented correlation between hopelessness and both the desire to die and suicidal ideation, even in patients without a life-limiting diagnosis (e.g. Beck et al., 1975).

The second virtue Navarini highlights is courage; if illness constantly requires the individual to reassess their circumstances and adapt to new challenges (including fear) then courage becomes essential. Courage, in this sense, enables the patient to face the uncertainties and constraints imposed by illness with determination and integrity.

Navarini concludes her triad by introducing practical wisdom, highlighting its deep connection with hope. She writes: "If hope is connected to agency, then deliberation regarding specific goals and the means to achieve them calls for the exercise of practical wisdom.

Hope regulates the emotions that motivate us to transcend today and the self, while practical wisdom orders concrete choices toward the good, thus offering, so to speak, the objects of hope" (Navarini, 2020, p. 135, my translation).

Practical wisdom, according to Navarini fulfills several functions: it enables creative adaptation to circumstances and guides the subject in identifying effective responses to everyday challenges, personal problems, and even moral dilemmas.

The philosopher does not specify how these traits might be cultivated in practice. However, she highlights on the hand the responsibility of the ill person to work on their character when possible. On the other hand, she suggests that it may be appropriate to introduce targeted interventions within early palliative care, that she refers to as "character care" that can promote the development of hope and wisdom[27].

Now that I have analyzed some of the most significant contemporary accounts addressing the compatibility between illness, suffering, and virtue, I will devote the final part of this work to examining the specific role of flourishing in the experience of illness. In particular, I will aim to identify those traits that can enable the patient to live a better life in illness. In this section, I will consider virtues not only as potential outcomes of positive coping with illness-related circumstances but as active drivers of change, i.e. resources that can help the ill person live more meaningfully and flourish.

27 For a more detailed discussion, see note 52 in Navarini's book (Navarini, 2020, p. 141).

PART THREE
THE PATIENT'S "ILLFULLNESS"
A CASE OF ETHICAL EXPERTISE

CHAPTER 5
THE EXPERT PATIENT

In the second part of this work, I began by analyzing the relationship between illness and flourishing, demonstrating that these concepts are in fact compatible. I focused in particular on one specific form of flourishing, namely, the kind of character development that can result of effective coping.

The final part of this work will explore a second approach to understanding the relationship between character and illness. In particular, I will investigate the protective function of flourishing and identify those character traits that may serve as important tools for helping individuals to live better within the experience of illness.

1. *Defining the "Virtues" of the Good Patient*

Before offering a classification of the virtues highlighted in the literature, it is essential to begin by clarifying their nature and understanding what is meant by the term "virtue"[1]. We can define virtues as positive character traits acquired through practice; once developed, they guide us in deliberating and acting in the most

1 To provide a definition, we could say, using the words of Campodonico, Croce and Vaccarezza that virtues are: "ethically positive character traits as opposed to the ones we consider negative (vices), acquired, i.e. the result of exercise guided by practical reason – itself directed toward the good end or human flourishing – and working in synergy with the emotions. Virtues are not habits or conditioned reflexes: they hence require the subject's freedom; they intrinsically possess a temporal dimension that presupposes the memory of the past, is acted out in the present and opens up in expectation to the hope of the future" (Campodonico et al., 2017, p. 87, my translation).

appropriate way. Virtues are relevant across all domains of life: we rely on them when deciding whether to give in to laziness and stay comfortably on the couch watching television, or to leave the house to help a friend in need; when a teacher chooses how to relate to their students, or a husband his wife; and we need them most of all when we face particularly dramatic moments and situations that evoke human frailty.

Following the articulation provided by Howard Curzer in *Is Caring a Virtue for Health Care Professionals,* we can divide virtues into two subtypes: "general virtues" and "role virtues". To describe these two classes of virtues, the philosopher writes:

> There are some moral rules which people generally should obey (Do not kill; do not steal; etc.). But different rules apply to people playing certain roles within the context of certain practices. (Soldiers generally should obey orders given by their commanding officers, but civilians have no such duty; parents generally should try to save their own children from a burning house before they try to save other children; etc.). The set of rules which applies to people playing certain roles is sometimes called a role morality.
>
> Role morality can be approached via virtue ethics, too. There are some character traits which, taken together, make someone a good person. A person lacking these virtues is morally deficient *qua* person. I shall call these general virtues. To be good at a role (to be a good lawyer or a good parent, for example), requires a collection of character traits which may be somewhat different from the general virtues. I shall call these character traits role virtues. I shall say with deliberate vagueness that role virtues are character traits which help the person achieve the goal(s) of the role (Curzer, 1993, pp. 51-52).

According to Curzer, *general virtues* would thus be the virtues *simpliciter,* i.e. global character traits or dispositions that, acquired over time and through practice, achieve a certain stability. They constitute, as Aristotle intended, a second nature for the subject that will enable them to activate the right behavior and to act well. For Curzer, general virtues are those traits that make a subject a good human being.

With the term *role virtues*, the philosopher refers instead to those dispositions that are developed by a certain person in relation to a certain role[2] (such as that of the soldier, or the commander) and that help them achieve the goal(s) associated with it. Curzer specifies that the virtues pertaining to a certain role do not always correspond with the general virtues; in fact, there may be cases in which a "general vice" also represents a role virtue (and vice versa).

To explain this, Curzer offers the example of the traits of competitiveness and benevolence: just as competitiveness represents an essential role virtue for businessmen, it is also a trait that could potentially make someone a bad person. Similarly, although benevolence is generally a virtue, it can also be a role vice. Let us think for example about a market strategist: in such a case it may be that too much benevolence may undermine their work, preventing them from achieving the desired results.

Following this overview, one may wonder whether the patients' virtues may also represent a case of role virtues. However, it is difficult to think of it in the same way as one would think of the role of a businessman, a lawyer, or a doctor. In fact, the lawyer, for example, can benefit from certain traits that can help them to better carry out their profession and achieve the goals connected with it, such as the ability to "nudge" a situation in a specific direction to defend the client's interests, even when the lawyer knows that the latter has committed a very serious crime. However, for the ill person it is different as there is no domain of life in which they cease to be ill; any severe illness, particularly when it involves chronic conditions and even more so in advanced or terminal stages, affects the patient's entire being. Illness conditions the subject's relationship with their own body and intimacy and the way in which they manage and interact with the spaces around them; it influences the way they relate with family members, with friends and how they interact in society; it conditions the way they manage their job, and so on.

Thus, we can say that a role is a "fluid" condition where, in a certain sense, one can dress or undress oneself – although some of

2 If we wanted to define what is meant by the term "role", we could say that it is characterized by a "set of behavioral expectations attached to a position in an organized set of social relationships" (Stryker, 2007, p. 1083).

the traits that compete with the role may end up permeating other domains of the subject's life. A role is not superimposable on the subject's identity as it represents just a domain of their existence. Precisely for this reason, illness cannot be regarded merely as a role. It is instead a condition that can permeate all domains of life.

To better illustrate this point, consider the example of Alex, a person who is naturally introverted. In her personal life, she avoids large gatherings and would never take the initiative to make new acquaintances at a party. However, in professional settings, Alex is required to behave differently. As a skilled salesperson, she knows the importance of networking and engaging with potential clients during business meetings. As such, she adopts an extroverted role in those contexts, even though it does not reflect her deeper temperament.

As is shown in the psychological literature, individuals can display significant variations in personality traits across different life contexts (Bleidorn, 2009; Heller et al., 2007). This does not imply that the agent lacks stable traits; rather, it reflects the influence that different social roles exert, often calling for distinct virtues or dispositions. For instance, the traits expected in the role of a mother can differ markedly from those required of a businesswoman (Heller et al., 2007). In this sense, virtuosity also demands the ability to adapt to changing circumstances and to shift between roles effectively. One of the hallmarks of a virtuous individual is, indeed, flexibility, that is, the capacity to attune behavior appropriately to the demands of varying situations (Bleidorn, 2009).

Returning to the notion of life in illness, it is important to clarify that being ill is not a role someone can get in and out of (unlike, for example, the patient role). Rather, it constitutes a condition (sometimes temporary, sometimes permanent) that permeates various dimensions of a person's existence. Because illness often has this pervasive influence, it is likely that the virtuous traits enabling an ill person to live well within illness are specific expressions or "shades" of the general virtues.

It is interesting to note that role virtues present challenges not only in the context of illness but also for virtue ethics in general. The concept risks endorsing the idea that someone might display virtues solely within a specific sphere (e.g. work), while being in general a bad human being. Yet, genuine possession of a virtuosity should imply consistency across various domains. This does not mean that identifying traits helpful in specific roles is unimportant, but it does raise a significant question: can one be truly virtuous in a role while failing to exemplify moral integrity in other areas of life? This theme is effectively addressed by the perspective that views virtues as clusters.

If role morality appears unsatisfactory, a more plausible alternative is proposal is to consider "virtues as clusters". To grasp its meaning, it is necessary to enter into the virtue ethicist reflection on the relationship between the virtues, i.e. whether, as Navarini suggests, there is a "solidary relationship" between them, meaning that virtues are interconnected and mutually reinforcing, or whether virtues possess a certain degree of independence from one another (see Navarini, 2022, p. 178).

There are basically three ways in which different authors have answered this question: the most classic answer is the "unitarist" view of the virtues. This first possibility can be subdivided into a "stronger" proposal such as the Socratic-Platonic and Stoic view which sees virtue as one and only one that can manifest itself by taking on distinctive characteristics. A "weak" option is, on the other hand, that of Aristotle, which portrayed the virtues as "an organic 'plexus' that leads the virtuous subject to gradually, but inescapably, possess the set of virtues coordinated by *phronesis*" (Navarini, 2022, p. 178, my translation).

According to the second view, it would be impossible to think about a virtuous subject's life as characterized by virtuous parts and vicious parts: a virtuous person should manifest the right virtue, the right behavior, and the right emotion in the right context. Moreover, according to this approach, being virtuous would also correspond to possessing all the virtues, directed by practical wisdom (the *auriga virtutum*), ensuring coherence and context-sensitivity. To better explain this concept, consider, for instance, the virtue of honesty: in a clinical setting, a doctor who is brutally honest may inflict harm, while one who is excessively evasive may obscure essential truths. A good doctor must therefore practice an appropriate degree of honesty, tailored by attentiveness, empathy, and tact. This nuanced judgment calls for practical wisdom; practical wisdom helps the subject understand which virtues are required in a given context and to what degree (Campodonico et al., 2017). Moreover, a virtuous person is not only expected to embody the qualities of a good doctor but also to demonstrate moral excellence across the various domains of their life.

In contrast to this idea, "disunitarists" argue that individuals can possess distinct virtues in certain areas while lacking them in others[3]. A person might be caring as a nurse but act harshly or unjustly in other domains.

Both visions have problems: a rigid unitarism may risk failing to account for subjective variations. Also, it seems to demand too much of the virtuous subject. As Mario De Caro and colleagues write:

> This thesis leads to a very unrealistic portrait of the virtuous agent as perfectly virtuous, and makes it impossible – again, unrealistically – to attribute genuine virtue to someone who is morally lacking in some specific respect. To put it differently, according to the unity of the virtues thesis, agents are either fully virtuous or utterly wicked since, in order to genuinely possess a single virtue, one needs to possess all the others (De Caro et al., 2020, p. 32).

Conversely, extreme disunitarism, such as that found in situationist views, struggles to account for consistent moral exemplars who reliably display virtue across various contexts[4].

A more nuanced alternative is the "virtues as clusters" model. According to this view, each context would represent a domain, which would invoke some particular virtues (or just one) correlated with one another. These clusters support and reinforce each other, making it more likely for a person to acquire the related traits (Gulliford & Roberts, 2018). Such a proposal seems particularly convincing as it suggests that each context elicits a unique constellation of virtues.

---

3 One answer in this sense is represented by situationism, which holds that there is no real "virtuous character" consisting of stable traits. In fact, according to situationists, it would be morally irrelevant traits/events (such as being in a hurry) that would decide the behavior of a certain subject in a certain situation, as demonstrated by the famous "Good Samaritan Experiment" by John M. Darley & Daniel Baston (for a review see e.g. Hauerwas, 2013).

4 On this see e.g. Zagzebski, 2017.

Aligned with this view, it is important to clarify that when I refer to "virtues", I do not mean isolated traits, but rather context-specific manifestations of practical wisdom. As I will elaborate further, following the Aretai model (De Caro et al., 2025), virtuosity is grounded in the possession of *phronesis*. Virtues are, therefore, *ratio conoscendi* of practical wisdom, i.e. its modes of appearance in different circumstances.

This way of understanding character appears to be quite realistic. Specifically, the vision I find most convincing is the one proposed by De Caro et al. (2021), which conceives of reality as divided into different domains, each of which calls for different manifestations of practical wisdom[5]. In order for one to be virtuous, the authors point out, one does not need to show virtuosity in every possible domain of experience since human beings are not ideal moral agents. On the contrary, they will need to possess a substantial degree of virtuosity[6]. The authors write about this:

> As we have argued extensively elsewhere (De Caro & Vaccarezza 2020), in order to see someone as moral, we must attribute a substantial degree of virtuosity to them – otherwise that agent would not look moral at all. Should we meet someone who excels in one moral domain (take respect for animals as an example) but is completely flawed in others (e.g. they exhibit cruelty toward some humans), we would not regard that person as virtuous at all. On the other hand, it is clear that our attribution of virtues to another agent does not have a holistic character, either. When we interpret other agents as moral, we never imagine that they possess all the virtues – we do not see them as saints (De Caro et al., 2020, p. 33).

---

5 The authors refer to Gulliford and Roberts' model.

6 It is crucial to clarify that the model proposed by De Caro and colleagues (2021), unlike other approaches, maintains that the only virtue truly possessed by a virtuous individual is practical wisdom, understood as ethical expertise. According to this view, there are no discrete virtues such as courage, patience, or hope; rather, these are expressions or manifestations of practical wisdom as it operates across various contexts. To be virtuous, then, is to exercise practical wisdom, which presents itself in different forms depending on the circumstance by requiring the subject to be courageous, to be honest, to be tactful, and so on.

In sense, according to De Caro and colleagues, mistakes do not disqualify someone from being considered virtuous; they simply reveal that the person is not a saint or a hero. Consequently, moral expertise should not be measured by perfection in every context or circumstance, especially in new or particularly challenging ones. Rather, the morally expert individual should demonstrate an "openness to new domains and situations", meaning an awareness of their own limitations and a willingness to grow (De Caro et al., 2021, p. 31). In this sense, even the wisest person is still capable of moral improvement (De Caro et al., 2018).

Returning to our cluster, that is, to the virtues specific to the patient's condition, we may think that the shades of "virtues" (i.e. of practical wisdom) acquired in healthy life serves as a reliable safeguard for the ill subject's moral character. This is true to a certain extent as a flourished character play a protective role. However, it may not be sufficient; it may even happen that a person who has always been considered a "virtuous exemplar" no longer appears to be so once they become ill. This could occur because the new context is significantly more demanding than any other situation they have previously faced. Thus, excellence can never be taken for granted, nor is it ever permanently secured.

Regarding the relationship between virtue and excellence, it is worth adding a few more reflections. When we think of a virtuous subject, we tend to imagine someone who has attained moral excellence[7] and who, as a result, consistently feels emotions and makes decisions aligned with that excellence of character. We would thus not expect them to suddenly stop being kind, patient, or courageous; if so, it would likely make us question whether they had ever truly been virtuous. This view, which suggests that a moral expert always behave virtuously regardless of the situation, not only reflects a well-established perspective in the literature – namely, that the behavior of a virtuous person is predictable across contexts – but also echoes the common tendency to envision virtuous individuals as closer to saints or superheroes than to ordinary people. Following

7 Some authors define excellence as the criterion for being virtuous (e.g. Adams, 2009).

this reasoning, if someone is known to be courageous, we will not expect to find areas or circumstances of life in which they are not so (even in the most complex ones). On the contrary, we expect their behavior to remain consistent, and for their courage to be ever-present. When this does not happen, we may feel disappointed and begin to doubt their virtuousness.

Yet reality often tells a different story. It is not uncommon for individuals we consider virtuous to struggle with especially difficult circumstances or to make mistakes. As De Caro and colleagues point out, while a certain degree of reliability is reasonable to expect from a moral expert, we cannot demand infallibility. As Matt Stichter notes, "we should not overestimate how well experts will react in novel situations, since their expertise is still linked to a certain history of experience" (Stichter, 2018a, p. 55).

There are indeed some areas of life in which even the wisest person may struggle to discern how to act. This is particularly true in situations that are exceptionally demanding, such as the experience of illness. To better illustrate the relationship between excellence and behavioral predictability, let us take a very different example and imagine a world without engines. In such a world, anyone needing to travel long distances must ride a bicycle. In this setting, there is a man named Charlie who is renowned as the best cyclist: he has always tackled even the most complex climbs, and is by far the fastest person in that world. Now imagine that, for some reason, all two-wheeled bicycles are suddenly banned, and people must use unicycles instead. Charlie would quickly discover that all of his previous skills are no longer sufficient. The challenges are now far more difficult. He struggles just to maintain balance, he's no longer the fastest, and he often has to dismount and walk uphill. With effort and dedication, Charlie may eventually learn to master the unicycle, but this would not happen automatically or without significant effort.

A similar dynamic can occur when someone faces a transformative experience such as illness. As we have seen, illness changes both the life as lived by the subject and the subject themselves. Consequently, things that once seemed trivial (like walking or waiting in line) can become major challenges. Likewise, concerns that once held great emotional weight may now appear insignificant, even leading to frustration or impatience.

If every aspect of a person's life changes, how can we expect them to flawlessly exercise the traits required by this new life? While having cultivated virtues in healthy life can certainly play a protective role and have a positive impact on the way the expert adapts to the new conditions, we should not expect this transition to be immediate. Experiencing illness demands not only a change in lifestyle but also a change in the person's character. It thus calls for new shades of practical wisdom that the individual may not have previously developed.

Indeed, while we might reasonably expect that virtuous agents will be better equipped to handle unfamiliar challenges, this expectation may falter when the challenges are particularly intense. As Lebacqz puts it: "To be a 'patient' [...] is to confront physical distress, personal threat, and a new social role. It is, indeed, a 'more onerous citizenship'" (Lebacqz, 1985, p. 278). Thus, it is impossible to fully predict how they will respond to such a new and burdensome condition. Some individuals may experience a true "epiphany", prompting a radical transformation in perspective and behavior. Others may initially undergo negative coping, leading to a weakening of character. As we will explore in the following paragraphs, it is not uncommon for someone facing a destabilizing new situation to react in ways that diverge sharply from their usual behavior, perhaps even adopting vicious habits, before eventually adjusting and returning to, or even surpassing, their previous level of moral competence.

Virtues must be "regained" through practice. What should therefore be expected of a moral expert is not perfection but rather awareness of their own limitations and the will to improve so as to "recapture" the virtuosity that now seems lost. In this sense, character education could be profoundly beneficial for the ill person. It could support their adaptation to new circumstances and even help them acquire the traits necessary to navigate life in illness more effectively.

It should be clear by now that becoming virtuous is a lifelong process that can undergo breakdowns. Moreover, it is important to note that this is not a linear process and that it is shaped by the subject's individual life path. In this sense, practicing virtuous traits even in new and difficult contexts makes a difference. In this respect, virtuosity appears to share many similarities with skills.

The theory that virtues are akin to skills has been widely explored in the literature. In the "virtue-as-a-skill" model proposed by Stichter, virtues are conceived as moral skills (Stichter, 2018a) that can be acquired through practice, much like any other skill[8]. Just as acquiring a practical skill involves learning by doing, character development is a process of experiential learning that leads to practical moral competence. Being virtuous, then, means being an expert in being good (that is, knowing how to perform the right actions, follow the right reasons, and feel the appropriate emotions), just as having a skill means being proficient in a particular area of practical life.

---

8 The virtue-as-a-skill thesis, which can be traced back to ancient Greek thought, has evolved into various formulations over time. Stichter (2018b, p. 58) classifies them into three broad categories: weak, moderate, and strong versions. In the weak version, exemplified, for instance, by Linda Zagzebski (1996), virtues are loosely connected to abilities but are not to be understood strictly as skills. The moderate view, represented most notably by Aristotle, posits a stronger analogy between virtues and skills. According to this perspective, virtues share important structural features with practical skills. As Stichter points out, Aristotle holds that there is a close resemblance between the moral knowledge of the virtuous person and the practical knowledge of the skilled expert. This analogy is strong enough that one might understand the development and structure of virtues by examining how skills are acquired. Nonetheless, Aristotle does not assert a strict identity between virtues and skills; rather, he maintains that their structures are analogous. The strong version of this thesis is represented by scholars such as Stichter himself and Julia Annas (e.g. 2011). Annas proposes that virtues should be understood as moral skills. This view does not imply that every skill qualifies as a virtue, but rather that the process of becoming virtuous resembles the process of acquiring expertise in a domain. The virtuous person, in this framework, is akin to a moral expert whose competence is developed through experience, practice, and reflection. This clearly does not mean that every ability is a virtue, rather, that those who are "experts" in a given field can be compared to virtuous persons.

Central to Stichter's model are experience, practice[9], and comparison with other moral agents. As he explains: "Virtues, like skills, require experience and practice to acquire. You cannot learn how to surf merely by reading a book about it, and likewise, you cannot acquire the virtue of kindness just by reading one of the current books on virtue. You need to learn by doing – to get good at surfing you need to surf, and acquiring kindness requires doing kind acts" (Stichter, 2018a, p. 2).

According to Stichter, the development of virtues demands "learning by doing." Crucial to this learning process is the combination of "routine practice" – repeating actions in contexts where the subject already has expertise – and "deliberate practice" – the exercise of one's traits in new, challenging situations (Stichter, 2018a, p. 24)[10]. Through such practice, the expert gradually learns to act effectively and intuitively in diverse contexts; indeed, what distinguishes experts from novices is their ability to intuitively recognize the right course of action in a given situation[11]. Stichter

9 Stichter writes on this: "The acquisition of virtue will require what is generally involved in acquiring skills. The most important factor is experience" (Stichter, 2018a, p. 64).

10 On the importance of practice for the expert see also Kahneman, 2011.

11 This aspect related to intuition has led to the emergence of a problem that has been referred to in the literature as "the automaticity challenge". This challenge states that if in exercising our skills we act without reference to conscious deliberative processes, we cannot say that such are rational skills (Sauer, 2012). Stichter responds to this critic explaining that when someone is an expert, they have internalized a kind of response pattern that allows them to respond to different types of situations without needing to have control over them. This characteristic of the expert subject – which also applies to the moral expert – has been referred to as *intended goal-dependent automaticity* (see Narvaez & Lapsley, 2005). Moral intuition in this sense would not demonstrate that the virtuous subject is irrational; rather, it shows that they have reached such a high level of expertise that they no longer need to rely on deliberation. This happens because experts have spent so much time engaging in practice in a given domain – Ericcson (1996) estimated, for example, that it takes about ten thousand hours of practice or 10 years to become an expert – that understanding what the right way is to respond becomes automatic. In this sense it is somewhat like when a pianist becomes an expert in playing a

writes: "Experts do not need to devote much conscious attention to what they are doing, and this lack of conscious attention does not lead to any reduction in their performance" (Stichter, 2018b, p. 64). Just as an expert driver instinctively knows when to accelerate or brake during a turn, so too does the moral expert intuitively know how to act.

This view characterizes distinguish Stichter's account from Julia Anns'. While Annas places significant emphasis on deliberation, Stichter sees deliberation as a sign that the subject has not yet reached full moral expertise. He also criticizes Annas's claim that a virtuous person must be able to articulate the reasoning behind their actions. Annas writes that the virtuous individual, unlike the non-expert, must "be able explicitly to explain and justify her particular decisions and judgements, and to do so in terms of some general grasp of the principles which define that skill" (Annas, 1995, p. 233).

Stichter, on the contrary, contends that moral experts are often unable to articulate their reasoning. To support this view, Stichter refers to one of Patricia Benner's studies on expert nurses, who despite their expertise, often struggled to explain why they acted a certain way. One of the nurses interviewed said: "When I say to a doctor, 'The patient is psychotic,' I do not always know how to legitimize that statement. But I am never wrong. Because I know psychosis from the inside out. And I feel that, and I know it, and I trust it" (Benner, 1994, p. 134).

Stichter, by contrast, argues that moral experts often cannot articulate the reasoning behind their actions. To illustrate his view, Stichter refers to one of Patricia Benner's studies on expert nurses. Despite their remarkable practical competence, these nurses frequently struggled to explain the logic of their clinical judgments[12]. One

---

certain piece: they have practiced for so long that their fingers move almost automatically. This does not make them irrational; it makes hand movements so stuck in their mind that they do not need to rely on deliberative intervention in order for them to be able to play the right notes, at the right time and in the right way.

12 In this sense, Stichter embraces a lighter version of the "Dreyfus model" of skill acquisition. The Dreyfus model argues for complete autonomy of choices and denies that the expert needs any deliberative process to make

particularly compelling example comes from an interview in which a nurse recounts: "When I say to a doctor, "The patient is psychotic", I do not always know how to legitimize that statement. But I am never wrong. Because I know psychosis from the inside out. And I feel that, and I know it, and I trust it" (Benner, 1994, p. 134).

This intuitive understanding of moral experts raises the question of the link between intuition and the ability to explain one's decisions. While a full exploration of this topic lies beyond the scope of the present discussion, it is worth noting that the literature (e.g. Stichter, 2018) suggests a possible connection between such intuitive competence and what has been described as the "flow experience". Mihaly Csikszentmihalyi (1991) defines flow as a state of deep immersion in an activity, where the individual loses self-consciousness and acts without needing to control their actions. The flow experience may help explain how experienced individuals can act competently without explicit deliberation. As in Benner's example, moral and practical knowledge may become so deeply

---

a decision. The Dreyfus model holds that the skill acquisition consists of five phases: novice, advanced beginner, competent performer, proficient performer, and expert (Dreyfus, 2004). Dreyfus writes about this: "It seems that beginners make judgments using strict rules and features, but that with talent and a great deal of involved experience the beginner develops into an expert who sees intuitively what to do without applying rules and making judgments at all. The intellectualist tradition has given an accurate description of the beginner and the expert facing an unfamiliar situation, but normally an expert does not *solve problems.* He does not *reason.* He does not even act deliberately. Rather he spontaneously does what has normally worked, and naturally, it normally works" (Dreyfus & Dreyfus, 1991, p. 235). According to this model, therefore, if the novice agent responds to situations by simply following context-free rules, what makes the expert as such is precisely the ability to detach themself from rules and maxims in order to make choices on the basis of intuition. The repeated exposure of the expert to a number of similar experiences would, over time, make them capable of taking immediate decisions. The only exceptions – which remain very rare according to Dreyfus – would be that in which they are confronted with new and extremely complex contexts. In these cases, even the expert should make use of deliberation to take the right choices.

ingrained through habit, experience, and reflection that it manifests as embodied and immediate.

Returning to the expert's capacity for intuition, Stichter argues that the best response often presents itself immediately to the expert because of past experiences[13] – and not as a result of a routine (or a pattern). Experts rely on their creativity to interpret situational elements and determine the appropriate action. In this regard, Navarini's example of the expert restorer is illuminating. Navarini explains that unlike novices, experts restorers can "sense" how to approach an antique artifact, even skipping or modifying procedures based on past experience (Navarini, 2020, p. 168). Similarly, the virtuous person intuitively knows what to do in various situations without needing to consult rules or weigh alternatives. Indeed, another feature of moral expertise, according to Stichter, is that virtuous actions are often performed effortlessly[14]. This is due to what he calls "goal-dependent automaticity", which depends on the subject's level of expertise. For instance, when a virtuous person sees their house on fire and knows their mother is inside, they automatically activate the goal "save mom" along with the corresponding virtuous response. This happens without conscious deliberation, reasoning about moral rules, or weighing alternatives; if this kind of response becomes frequent, the goal becomes "chronically accessible" and thus easy to activate (Stichter, 2018a, p. 68).

It is important to add that chronically accessible goals are not, as Snow notes "reflex reactions to stimuli, but [...] intelligent, flexible responses to unfolding situational cues" (Snow, 2010, p. 43). Repeated exposure to situational cues reinforces these goal-directed behaviors, making them more automatic yet not mindless. As Stichter emphasizes, this kind of behavior becomes automatic "without it being unthinking or simple rote behavior" (Stichter, 2018b, p. 65).

13 According to Stichter, the expert's ability to intuit how to act in a given situation would not only come from practice (which undoubtedly plays a fundamental role), but also through the use of mental heuristics. Stichter defines mental heuristics as: "shortcuts in reasoning, where you simplify a complex problem in order to come to a decision more easily" (Stichter, 2018b, p. 65).

14 On this see also Feltovich et al., 2018.

It is important to note that even experts cannot always rely on intuitions. According to Stichter, certain situations – those that are complex, unusual, or morally dilemmatic – still require conscious deliberation. Stichter writes regarding this: "Virtue is often discussed as something that once you acquire it, then it can guide you appropriately in whatever situation you face (and hence the claims to cross-situational consistency). But unusual problems or difficult conditions must pose challenges that it will take some experience and practice to figure out how to deal with, otherwise they would not be unusual or difficult" (Stichter, 2018a, pp. 178-179).

Thus, when confronted with unfamiliar or particularly challenging situations, even a moral expert may be forced to rely on deliberation. As previously noted, Stichter, like De Caro and colleagues, does not portray moral experts as infallible beings[15].

15 Although it is possible to see many similarities between virtues and skills, there is one element that casts doubt on this theory, namely, the problem of *motivation*. As Linda Zagzebski, for example, observes in this regard, while to say of a subject that they are an expert it is not necessary for them to show the possession of a certain skill on every occasion and in any context, moral expert needs motivation to act. The virtuous subject cannot in this sense, for example, be aware that they can help a friend, know that helping them can make a difference, and not being motivated to do so. However, according to Zagzebski, this is different, for an athlete who can decide in a certain context not to show their skill; a skill in this sense does not stop being so if it is not exercised, and this says nothing about how skilled one is. On the contrary, it seems that the virtuous subject should always be motivated to act well. However, the idea that skills do not need the element of motivation seems to be unconvincing. Indeed, as Gary Watson points out, judging how skilled a person refers not only to their ability, but also to when and how much they are motivated and committed in a certain practice (Watson, 2004). If in fact an expert dancer did not feel motivated to give their best in a performance this would seem odd. This point is also shared by Darcia Narvaez, who writes that experts desire excellence in their field (see Narvaez, 2006, p. 719); the practice of any expert therefore cannot do without the element of motivation.

2. *Descending Transformativity and Ascending Transformativity*

Before delving into the core discussion of patient virtues, I would like to spend a few more words about what happens to the moral expert who encounters difficult experiences such as that of illness. Indeed, as I previously noted, when confronted with particularly complex circumstances, even a virtuous individual may feel as though they have lost the virtues they had previously acquired. However, over time, they may develop a new context-specific nuance of the virtuous trait in question. I refer to these processes as "descending transformativity" and "ascending transformativity".

In circumstances such as illness, transformativity can have different outcomes, leading either to a strengthening of the character – though this must pass through a critical phase – or to a weakening of it. When a subject is confronted with the communication of an inauspicious diagnosis, they will almost inevitably find themselves at the center of a profound crisis. This crisis often constitutes a necessary step within the coping process. If it unfolds constructively, the individual may achieve a positive adaptation to the illness, which can result in PTG and, consequently, to the development or consolidation of positive character traits and virtues. In such cases, the individual may undergo what could be described as an epiphany[16] which can be defined as a shift in the structure of meaning they previously attributed to their life.

Conversely, certain experiences may be so deeply traumatic that, if left unsupported, they can provoke a gradual erosion of the subject's character, particularly when coping mechanisms fail. Dysfunctional coping may occur for different reasons. This may happen when the subject is unable to reframe their condition in a meaningful or manageable way, perceiving it instead as irredeemably unjust and intolerable. In such contexts, a well-structured palliative care pathway can play a pivotal role in assisting the subject to accept their new condition and to re-signify it in a more constructive light.

16 By the term "epiphany" I refer to what Norman Denzin defines as: "interactional moments and experiences which leave marks on people's lives. In them, personal character is manifested. They are often moments of crisis. They alter the fundamental meaning structures in a person's life" (Denzin, 1989, p. 70).

This, in turn, can foster the development of those character strengths and virtuous traits that, as the literature indicates, may significantly influence the patient's response to illness, ultimately enhancing both their quality of life and overall well-being. As Kidd notes, "an ill person who is courageous and patient will likely be better able to cope with their illness and its treatment" (Kidd, 2012, p. 503).

However, more frequently, what is observed in those who undergo such traumatic experiences is a seeming regression of their previously developed traits. This is what I refer to as *descending transformativity*. This regression may in some cases be permanent and lead to a gradual withering of character. But just like character development, this too unfolds within a process. While it is certainly true that virtues can be lost one small piece at a time, similarly, there is hardly ever a clean break or a sudden change. What typically happens is that, when confronted with a new and complex situation, the person needs time (and practice) to adjust. The initial shock and the negative stages typical of the coping process may produce a temporary loss of virtuosity. Yet this loss is not always irreversible.

Indeed, it often happen that the subject, by positively adapting to the new condition, may not only recover their previous moral competence but even develop a higher level of moral expertise. This is what I define as *ascending transformativity*. In this case, the subject will have developed new nuances of the same virtuous traits, namely, of practical wisdom as it manifests in that particular domain.

To better understand this dynamic, consider the case of a young athlete who has consistently placed first in their category. If they move up to compete in a new category, they may possess the necessary skills, but it would be unrealistic to expect them to win their first race. They would need to train harder, learn new strategies from experts, and adjust to new equipment before they could gradually reclaim the status of champion.

This analogy helps to understand that the experience of illness often marks a new chapter in life, a "before" and "after" to that which calls for a reconfiguration of one's character. However, this should not lead to the mistaken belief that an ill person who was previously regarded as virtuous has lost all their virtues, or that they were never truly virtuous. Facing transformative and particularly demanding circumstances requires renewed effort, training, and moral exercise. However, through this process, character can be "regained" and, as the literature shows, sometimes even strengthened.

Therefore, before concluding this chapter, it is crucial to emphasize that the virtues needed for life in illness do not perfectly mirror those cultivated in healthy life. As we have seen, virtues are not rigid or monolithic entities, but rather resemble archipelagos, composed of multiple nuances of the same trait. In this sense, the general category of courage includes battlefield courage, parental courage, surgical courage, each of which is a context-specific instantiation of practical wisdom activated in different settings[17]. If this view holds, then exercising a virtuous trait becomes more difficult in the context of illness not only because the situation is inherently more demanding, but also because the person is navigating unfamiliar terrain. They may lack the specific nuance of practical wisdom required, and thus have to acquire it gradually, through repeated exposure and moral practice. This is the essence of a transformative experience.

Now that we have explored the complex relationship between virtue, suffering, and flourishing, I will turn to the identification of the character traits that may form a core cluster of virtues necessary to live well in illness.

Up to this point, I have examined how character growth can emerge as a result of a positive response to illness. However, as noted earlier, a virtuous character may also serve as a protective factor that facilitates such adaptation, even though this should never be taken for granted. This hypothesis is supported by several empirical studies (e.g. Campbell & Swift, 2002; Dekkers et al., 2004; Ruini & Vescovelli, 2013; Swift et al., 2002). For example, the study by Swift and colleagues (2002) shows the role that certain virtues can play in the experience of illness which clearly emerges from patients' narratives. One of the participants remarks: "it sounds awful to say it but – a certain strength of character, I must have, otherwise I couldn't have coped with some of the things that have happened in my life" (Swift et al., 2002, p. 476). This

17 Each virtue cluster is thus composed not of general virtues, but of shades of the general virtues.

testimony highlights that virtues are not only potential outcomes of illness, but may also serve as essential resources for confronting it.

Wright and colleagues echo this idea: As they write: "[F]lourishing requires the development of robust dispositions to respond well to difficulties, setbacks, trials, and tribulations. And responding well requires a balance between strength and vulnerability – being able to tackle problems and handle difficulties on our own, but at the same time to support, and be supported by, others through these difficulties" (Wright et al., 2020, p. 60).

If, then, a flourishing character can indeed serve as a protective factor, it makes sense to reflect on which specific traits are most conducive to this end. In particular, I will distinguish between two set virtuous traits, namely, "self-regarding" virtues and "relational virtues". While the first set concerns how the subject relates to the illness and its challenges, the second concerns the traits that facilitate the patient's relationship with others. Since caregiving is a practical, relational activity involving both the carer and the cared-for, fostering those traits that enhance relational quality may be particularly important in the context of illness.

It is important to note that the aim of this section is not merely to identify the protective factors that may help a patient face the experience of illness and navigate its challenges. Rather, this is intended as a preliminary attempt to identify the kinds of inner resources that ought to be actively supported and cultivated through dedicated character education pathways, starting from the moment of diagnosis. As we have seen throughout the previous sections, the experience of illness should not simply be approached with the assumption that the patient already possesses a flourishing character. On the contrary, the vulnerability and disruption introduced by illness often reveal the need for moral support and guidance. Precisely for this reason, healthcare facilities should consider the implementation of protocols that support the development or reinforcement of specific virtuous traits. These would not only help the subject to cope more effectively with their condition, but also promote a better quality of life in illness, allowing them to continue pursuing meaning despite the limitations imposed by their health.

3. *Self-regarding virtues*

3.1. *Hope*

The first set of traits I will focus on comprises self-regarding virtues. My discussion begins with the virtue of hope.

Hope is a trait whose nature remains the subject of ongoing debate in the literature. Philosophers such as the Stoics, Descartes, Hobbes, and Hume have traditionally understood hope as an emotion or passion. Thomas Aquinas was the first to assign it the status of a virtue, though he classified it as a theological virtue[18]. Only more recently has hope begun to be recognized as a human virtue[19].

Hope has often been confused with related constructs, most notably, optimism. It is important to clarify from the outset that these two traits are not equivalent. As Navarini points out, optimism is grounded in confidence regarding the likelihood of achieving a certain outcome, whereas hope is broader and remains operative even (and I would add especially) when the desired outcome is uncertain or unattainable[20].

Snow further illuminates this distinction by identifying two essential elements that differentiate the two concepts. The philosopher notes that although both hope and optimism can be regarded as dispositions, they differ in their underlying attitudes: hope is characterized by anticipation, while optimism is marked by expectation. The hopeful person looks forward to a possible future without certainty, aware that obstacles may arise. The optimist, by contrast, assumes that the desired outcome will come to pass:

---

18 Aquinas also distinguishes hope from desires. As Mittleman writes in discussing the difference between desire and hope in Aquinas: "One might desire to sprout wings and fly away but, in Aquinas's terms, one cannot hope for this. Hope entails a realistic appraisal of possibility. Desire does not" (Mittleman, 2009, p. 43). Hope, according to Aquinas, is a passion that stimulates us to reach what we are hoping for.

19 See for example Navarini, 2020; 2022 and Snow, 2013; 2018; 2020.

20 To further explore the difference between optimism and hope see Eagleton, 2015.

> Optimism is characterized by expectation, which involves (defeasible) confidence that the expected outcome will occur (the confidence is defeated when the expected outcome does not occur); whereas hope is characterized by anticipation, which involves (defeasible) doubt (though not necessarily suspense) that the anticipated outcome will occur (the doubt is defeated when the anticipated outcome occurs). More simply put, we can say that optimism is characterized by confidence in a desired outcome's occurrence, whereas hope is colored by doubt (Snow, forthcoming).

As Navarini observes, optimism can be considered a form of "pseudo-hope", a more fragile trait that often collapses when the desired outcome fails to materialize. Hope, in contrast, drives the subject to reorient themselves toward new goals or alternative pathways (see Navarini, 2020, p. 194).

Having clarified these conceptual differences, let us turn more deeply to the nature of hope itself. Alan Mittleman, in a monograph entirely devoted to hope, argues that it is a trait peculiar to human beings; animals lack in fact, according to the author, the ability to imagine the future. Mittleman describes hope as a fundamental disposition with significant "adaptive survival benefits" (Mittleman, 2009, p. 25). To explain this further, the philosopher refers to the writings of Jerome Groopman, an American oncologist, who dedicated much of his work to explaining the importance of hope in severe illnesses like cancer. In his *Anatomy of Hope,* Groopman shows how patients with a greater capacity for hope tend to respond better to treatment and experience improved quality of life. His reflections echo Viktor Frankl's (1984) earlier findings; Frankl observed that concentration camp prisoners who survived were often those who maintained a sense of hope. Indeed, hope has not only a symbolic role in benefiting the subject who experiences it, it seems to also have biochemical one. Groopman explains:

> I found that there is an authentic biology of hope. [...] Researchers are learning that a change in mind-set has the power to alter neurochemistry. Belief and expectation – the key elements of hope – can block pain by releasing the brain's endorphins and enkephalins, mimicking the effects of morphine. During the course of an illness, then, hope can be imagined as a domino effect, a chain reaction in which each link makes

improvement more likely. It changes us profoundly in spirit and in body (Groopman, 2004, p. xvi).

While Groopman's account is valuable in highlighting the therapeutic role of hope, it is not without issues. Specifically, his view presents hope as a passive emotional response, something that "happens to us" rather than something we cultivate or practice. This position is problematic for two reasons. First, it assumes (i) that emotions are beyond the subject's control. Second, (ii) it treats hope as an emotion rather than a moral virtue.

On point (i), there is substantial evidence to say that emotions can be shaped, trained, and regulated by a virtuous character. As Gross (2002) defines it, emotion regulation refers to "the processes by which individuals influence which emotions they have, when they have them, and how they experience and express these emotions" (p. 275). Authors such as Paul Carron (2014), Kristjánsson (2018), and Stichter (2020) have shown that moral agents can gain control over their emotional responses through deliberate moral practice.

On point (ii), the classification of hope as a mere emotion is unconvincing for reasons articulated by Mittleman himself. Hope, he argues, is not simply a feeling as it plays a regulative and guiding role in a person's inner life. Indeed, the author notes that hope cannot be counted among emotions since it has a priority over them, just as it also has a priority over feelings and desires. In fact, while a person may be overcomed by fear, frustration, or despair, hope allows them to reframe these experiences, to mute the internal noise, and to continue pursuing what matters.

Navarini offers further support for this view. She argues that hope can be classified as a moral virtue precisely because of its ability to regulate emotions (typical of moral virtues) and for its capacity to motivate the subject to act (see Navarini, 2020, p. 128). To make an example, a hopeful patient is not immune to negative emotions like fear or anger; rather, they are able to manage those emotions, continuing to strive toward meaningful goals despite hardship. In this sense, hope is not just a psychological state, it is a moral disposition.

Having clarified some compelling reasons why hope should be considered a virtue rather than an emotion, let us now turn to providing a clear definition of it. A particularly convincing definition comes from the work of psychologist Charles Snyder, who defines it as the combination of two essential components: the willingness to pursue new goals (*agency*) and the knowledge of how to achieve them, and the ability to change one's path several times (*multiple pathways*)[21].

The relevance of hope in the context of illness becomes immediately apparent when viewed through the lens of Snyder's framework. As discussed in the early chapters, illness can profoundly disrupt a person's life. Everything previously taken for granted (bodily integrity, routines, expectations, dreams, and relationships) can be altered by the onset of a serious illness. It is precisely in this sense that illness can be defined as a transformative experience, drawing a sharp line between a "before" and an "after".

To better understand the value of this trait in this context, let us return to an earlier example from chapter 3.4, the case of Carl, a young athlete diagnosed with an aggressive form of osteosarcoma, who is told that the only chance for survival lies in the amputation of both legs. While Carl understands that this procedure is medically necessary, he also knows it will completely reshape his life. The surgery will permanently alter his body, limit his mobility, and shatter his dream of becoming a professional athlete. This moment signifies a rupture in Carl's narrative identity, changing not only his physical reality, but also the trajectory of his goals and the meaning he assigns to his future.

The way Carl adapts to this experience will be deeply influenced by his ability to integrate this change meaningfully into his life. Hope, understood as the ability to pursue alternative goals, to visualize new ways forward, and to remain open to revising those pathways, can help Carl reclaim a sense of agency and direction. If cultivated, hope may help him rediscover purpose and meaning. Hope, in this sense, is a crucial resource in the experience of illness: it prevents

---

21 Indeed, Snyder defines hope as "a positive motivational state that is based on an interactively derived sense of successful (a) agency (goal-directed energy), and (b) pathways (planning to meet goals)" (Snyder et al., 1991, p. 287).

the subject from giving in to despair, encourages them to trust that new opportunities will emerge, and supports their resilience in the face of ongoing challenges.

As Snow (2018) recalls, hope is also fundamental in helping the subject to become *resilient*. She writes about this: "A despairing person, one fearful of not attaining a hoped-for end, or a cynic, would lack the motivation and cognitive resolve to be resilient or persevere in the face of obstacles or threats. Being resilient requires strength. People lacking in hope seem not to have that strength. By contrast, someone who is hopeful in the face of obstacles or threats is given strength by their hope" (Snow, 2018, p. 16). For this reason, in one of her more recent works, Snow identifies hope as one of the foundational "building blocks" of resilience (Snow, 2025)[22]. Indeed, life in illness is often marked by moments of profound complexity and discouragement. Daily tasks may become disproportionately difficult; treatments may prove ineffective; and physiotherapy may not yield the expected improvements. Over time, such setbacks can generate deep frustration and may ultimately lead the patient to experience hopelessness which has been shown to correlate strongly with depression, a desire to die, and increased suicide risk (see Beck, 1986; Beck et al., 1975). In this context, cultivating the virtue of hope can have a transformative impact, potentially altering the course of the ill person's life.

This, of course, does not mean that the ill person should cling to unrealistic desires, such as the immediate recovery from metastatic cancer. Hope qualifies as a virtue only when it is oriented toward goals that are intelligible and attainable. It must be thus guided and governed by reason. Conversely, hope that is detached from reality (hope for the impossible) constitutes false hope, which offers no real benefit to the patient. At best, it may result in disappointment; at worst, it can lead to misguided decisions. As Groopman observes:

---

22 While Snow acknowledges that resilience can support the emergence of hope, she also maintains that neither trait independently guarantees the attainment of the desired outcome. Hope and resilience, though mutually supportive, are each insufficient on their own. For this reason, she considers them necessary but not sufficient elements in the process of coping and moral flourishing.

"False hope does not recognize the risks and dangers that true hope does. False hope can lead to intemperate choices and flawed decision making. True hope takes into account the real threats that exist and seeks to navigate the best path around them" (Groopman, 2004, p. 198).

It follows that any character education pathway must also include guidance on directing hope toward appropriate and reasoned ends. Only in doing so can the patient experience what we might call true illfullness. I will return to this point later.

3.2. *Courage and Perseverance*

It should now be clear that hope is a necessary trait for the ill subject, as it supports them in facing the many obstacles and challenges that illness and its consequences bring. However, hope alone is not enough. Other virtues are required to sustain the patient. Consider, for example, a person with advanced cancer: they will likely undergo multiple rounds of chemotherapy and experience a range of adverse side effects; they may begin pain management therapies that involve medications with unforeseen consequences; they might need to undergo a series of complex and potentially life-threatening surgeries. In such a context, it becomes essential for the patient to cultivate dispositions that will enable them to pursue their therapeutic goals and confront the fears that arise along the way. In this regard, as Navarini notes, hope "invokes" the virtue of courage (Navarini, 2020).

Aristotle describes courage as an essential trait for facing adversities, especially when the subject has inadequate reasons for expecting a certain kind of outcome. Courage is very important for patients as it can foster a constructive response to the many fears and uncertainties imposed by illness such as the fear of the diagnosis' communication at each follow-up; the fear that radiation or chemotherapy will fail; the fear of no longer recognizing oneself due to bodily changes; the fear of losing loved ones; the fear of dying, and so on. Moreover, courage becomes crucial in confronting the inherent risks of treatment. Many medical interventions, such as certain medications or invasive therapies, involve significant uncertainty and potential harm. When beginning a new therapy, the patient may have no guarantee of recovery or even of improvement. At best, the treatment might slow the progression of the disease or prevent further complications. The patient must commit to taking medication, undergoing treatments, or consenting to surgery without knowing whether any of it will prove effective. They may never wake up from the surgery; or, upon waking, they may be told that no curative option remains. They may endure exhausting side effects from treatments that ultimately fail to produce any tangible benefit.

For these very reasons, there are many authors who stress the importance of courage for the experience of illness (e.g. Brady, 2018; Carel, 2016; Kidd, 2012; Navarini, 2020; 2022; Shelp, 1984; Snow, 2018). Brady in particular includes it among the "virtues of strenghts", namely, those virtues that provide the ill person with the "'disposition to stay firm' against threats and harms and enable strong and reliable governance of our feelings and actions" (Brady, 2018, p. 95). The virtue of courage, sometimes referred to as fortitude, is fundamental according to Brady, as it encourages the ill person not to succumb to complications and to overcome the obstacles that illness often imposes. Courage is essential because it equips patients with the necessary strength to take decisive action: where hope identifies alternative paths, courage provides the tools to set fear aside. Think in this sense, for example, of a patient who has to undergo open-heart surgery to recover from an illness; remaining in a contemplative dimension hoping to recover is not enough. It is the decision to do preoperative examinations, to get on the waiting list, to sign the informed consent and to get on the stretcher that will help them realize their hope. Courage, in fact, works on the patient's fears and anxieties, generating the inner strength required to take certain actions and make certain choices.

As Navarini writes, courage can also "be a protective factor with respect to the anguish typically associated with certain stages of illness and especially the foreshadowing of the darkest scenarios" (Navarini, 2022, p. 197, my translation). This does not mean that the courageous patient will no longer experience emotions such as fear. Contrary to what one might assume, fear is crucial, not only because courage would not exist without it, but also because it has a regulatory role. It is precisely because a person perceives a potential danger that they can avoid taking reckless actions (see Navarini, 2020, p. 132). What distinguishes the courageous person is not the absence of fear, but the ability to manage it, containing its disruptive effects so that it does not hinder decision-making or paralyze action (Navarini & De Monte, 2019).

It is important to note that for courage to be virtuous, it must not degenerate into recklessness, as this could lead individuals to make dangerous or even fatal choices. As with hope, courage can only be considered a virtuous trait if it is governed by reason. For instance, a firefighter who knowingly enters a collapsing building with no realistic chance of saving anyone is not displaying courage but rather recklessness. In this regard, we return to the key function of emotional regulation that defines moral virtues: moral virtues are capable of tempering emotions that could otherwise lead a person to act in excess or deficiency, depending on the context. The courageous person is therefore not the one who feels no fear or who rushes into action without forethought, but rather the one who is able to set aside overwhelming or paralyzing emotions to achieve a greater good.

When talking about courage, the importance of the virtue of hope returns in that, as Navarini writes, "the courageous person must be clear about both the purpose of the action [...] and the concrete possibility of achieving it" (Navarini, 2022, p. 194, my translation); in order for someone to be courageous, they need to believe that the goal is attainable.

Courage proves thus essential for the ill person, who is constantly confronted with complex circumstances and perceived threats. For example, many medical treatments require extended periods before producing tangible benefits, and medications must often be taken regularly and consistently, even when, in the early stages, they seem to yield only negative side effects or appear entirely ineffective. Similarly, patients are often required to engage in daily physical exercises and attend regular physiotherapy sessions in order for these interventions to have any real impact. They may also be asked to modify their diet or adopt new lifestyle habits. These are just a few examples of the routine demands placed on the ill person and require not only the activation of the virtue of hope in setting and committing to meaningful goals, as well as the courage in pursuing them despite uncertainty and fear. They also include the virtue of *perseverance* which is essential to sustaining effort over time and enduring the repeated challenges that illness entails.

Perseverance has often been juxtaposed with the virtue of courage or included within the broader cardinal virtue of fortitude (e.g. Brady, 2018; Navarini, 2020; 2022). Navarini, for example argues that holding these two traits together helps to understand the two constitutive dimensions of courage, namely, the more operational part, i.e. the "attack" component, and the component of "endurance". It is the latter which involves perseverance. Navarini explains what is meant by the two expressions through two particularly useful examples: the so-called "courage of attack" is what Paul may need when giving himself an injection, if he has always been afraid of needles. "Courage of endurance" can be exemplified by a scenario in which Paul gradually learns to manage his emotions and withstand the discomfort he previously experienced during various withdrawals and injections (see Navarini, 2020, p. 134). Courage of endurance comes into play in the illness journey since disease-related goals are generally long-term goals. As Navarini writes, in fact, endurance allows the subject "to continue pursuing goals when the future looks different from what one imagined, and in general allows them to continue pursuing their own idea of 'good', despite the discomforts that illness brings" (Navarini, 2020, p. 114, my translation).

Although the virtues of courage and perseverance are clearly interrelated and mutually referential, I argue that they are best treated as distinct traits. Differentiating them allows us to better identify the specific functions and demands associated with each. Consider, for instance, a patient facing an invasive surgical intervention: in that case, it is not perseverance they require (which may come into play during postoperative recovery), but rather the kind of inner strength that enables to confront fear, namely, courage. Conversely, a person beginning a long physiotherapy regimen to recover motor function is not necessarily called upon to act courageously (unless the therapy entails some specific risk), but rather to cultivate the kind of tenacity that prevents them from giving up in the face of slow progress or repeated setbacks. In that context, perseverance is the relevant virtue.

Having clarified why it may be appropriate to keep the two traits conceptually distinct, we may now ask what makes perseverance such a crucial virtue for the ill person. Perseverance is particularly relevant in the context of illness because it supports the pursuit of long-term health-related goals which are often difficult to achieve. In this light, perseverance serves as a motivational anchor, helping patients maintain consistency in their efforts. It sustains the discipline needed for daily medication adherence, compliance with treatment protocols, and ongoing engagement with rehabilitation plans. More broadly, it instills the grit and resilience necessary to face discouragement and setbacks without losing sight of one's objectives (Dekkers et al., 2005).

Indeed, life in illness often resembles a marathon: progress is slow, energy may wane, and goals can seem distant. It is precisely perseverance that provides the inner stamina required to stay the course. As such, perseverance plays a key role in enabling certain acts of courage and also complements the function of hope. This interdependence among virtues underscores the need to think of them as forming a unified whole. Still, perseverance can and should be recognized as an autonomous trait

3.3. *Humor*

Another trait that plays an important role in the context of illness is humor. The inclusion of humor (explicitly defined as such) among the virtues is a relatively recent. Frankfurt (2001), for instance, presents it as a resource capable of supporting individuals throughout life, not by denying imperfections or limitations, but by assigning them appropriate weight. This conceptualization has proven particularly influential. On one hand, humor is often described as a disposition eliciting joy and laughter in others, especially through the perception of incongruities, paradoxes, or unexpected perspectives on reality (Dionigi & Gremigni, 2010, p. 27). On the other hand, it is also understood as the capacity to smile (or even laugh) in the face of adversity (Beermann & Ruch, 2009).

Not all scholars agree on humor's status as a virtuous trait. Julia Annas (2011), for example, argues that humor cannot be considered a virtue, since it can be used for both morally good and morally harmful purposes. Yet this objection does not seem decisive, as the same can be said of many other virtues. A virtue may become a vice if misapplied or pursued in excess, and even a morally admirable traits can have negative effects when used in the wrong context. Hope, for example, may become a negative trait if it is placed in unrealistic outcomes and pursued through inappropriate means.

In this regard, humor may be seen as a trait that can be directed toward good or bad ends. As Lott points out, "[h]umor can be clearly used to harm, encourage, criticize, console, etc. If I tell a joke in order to humiliate my colleague, or to denigrate a class of people, and if I act with that purpose in the mind, then I cannot escape moral responsibility by insisting: 'It was only a joke!'" (Lott, 2021, p. 116). Not every form of humor, then, is virtuous. As Alfano and colleagues emphasize, humor becomes a virtue only when it is exercised "about the right things, at the right time, in the right way, with the right people, for the right reason, with the right intensity, and so on" (Alfano et al., 2022, p. 662). This refined form of humor can be referred to as "virtuous humor" and can be considered a moral virtue precisely because of its emotional regulatory function.

Speaking about the role of humor in the experience of illness, it may serve as valuable resource on both an individual and relational level. On the individual level, humor helps regulate emotions and contributes to the re-signification of negative experiences. In fact, it facilitates more effective coping by opening access to complex emotional states and allowing individuals to approach painful or otherwise inaccessible topics. As Fotini Kounti-Zafeiropoulou (2014) points out, it does this by producing "cognitive shifts" that enable the subject to reinterpret their circumstances from a more constructive or positive perspective.

Furthermore, numerous studies have shown that humor correlates positively with self-efficacy, a renewed sense of purpose, increased self-esteem, and greater acceptance of illness and disability. Conversely, it is negatively associated with psychological distress, stress, and depression (e.g. Eleni et al., 2024). Furthermore, humor appears to be closely tied to hope and may even help to activate it (Karami et al., 2018).

Humor finally has a relational dimension, as it may facilitate communication between patients and caregivers, strengthen interpersonal bonds, and contribute to fostering a more positive emotional atmosphere (Wanzer et al., 2005). For all these reasons, it should be recognized as a potentially virtuous trait within the context of illness[23].

3.4. *Patience*

Now that we have analyzed the virtues of hope, courage, perseverance, and humor, we can now turn to examine the final two traits that complete the cluster of self-regarding virtues: *patience* and *humility*. I will begin with the former.

To define patience, we may refer to the work of Sarah Schnitker and colleague, who describe it as "the propensity of a person to wait calmly in the face of frustration, adversity, or suffering" (Schnitker et al., 2017, p. 263).

Although this trait can certainly be categorized as a self-regarding virtue, it is more accurately understood as a hybrid trait. As Lebacqz writes, patience concerns "how we stand in relation to other realms of being" (Lebacqz, 1992, p. 291). In this sense, while patience undoubtedly relates to the individual's regulation of their internal world, it also involves their orientation toward others. Indeed, during the course of illness, the patient is called not only to accept the limitations imposed by fear and anxiety or to overcome the physical and emotional obstacles that illness entails. They are also required to sustain hope in the eventual realization of meaningful goals, and to persist with treatments even when these do not immediately yield the desired outcomes. Illness, moreover, entails waiting for treatments to work, waiting to rediscover a sense of self, to feel at home in one's body again, or to rebuild relationships that may have been affected. The virtue needed to navigate these prolonged and often uncertain periods is, unsurprisingly, patience.

---

23 For a more in-depth exploration of the virtue of humor in the context of illness, see Ricci, 2025.

These examples reflect the self-regarding dimension of patience, which, like perseverance, is rooted in endurance. This is likely why Brady includes it in his cluster of "virtues of strength," noting its ability to regulate emotional states[24]. Indeed, patience often undergirds perseverance, as maintaining long-term efforts frequently depends on the ability to exercise self-restraint and emotional stability.

It is precisely this capacity for self-control that enables patience to function as a bridge between self-regarding and relational virtues. Patience, in fact, not only supports the ill person in pursuing their goals and managing their own frustrations; it also plays a crucial interpersonal role. Small, everyday challenges can easily become sources of distress. Patience is thus vital not only for gradually learning to accept one's new identity and waiting for treatments or therapies to take effect, but also, as shown in the literature (e.g. Schnitker, 2012; Schnitker et al., 2017), for regulating levels of frustration and promoting psychological well-being. As Denise Vigani points out, patience involves not just "waiting patiently," but also "acting patiently" (Vigani, 2017, p. 329). This distinction is essential, as it highlights that patience is not a passive trait but an active stance toward adversity. By helping individuals avoid venting their frustration or anger through impulsive gestures or words, patience can protect the quality of their relationships, whether with family members, caregivers, friends, or healthcare professionals. It is precisely in this regard that patience's relational dimension clearly emerge.

3.5. *Humility*

To conclude our discussion of self-regarding virtues, I will now turn to the analysis of humility. Like patience, humility also has a strong relational component. It thus serves as a critical bridge between the self-regarding virtues and the relational virtues.

Humility, which we may define in Lebacqz's terms as the "sense of one's limits" (Lebacqz, 1992, p. 292), is particularly important in the context of illness, as it enables patients to recognize their dependence

---

24 He draws this element from Hurka, 2000.

on others and opens them to the dimension of care. This does not entail self-abnegation, as some critics, including Hume, have suggested by portraying humility as the acknowledgment of one's "deformities" (Hume, 1960, p. 286). On the contrary, it is a moral disposition that allows the agent to accept their limitations and to become aware of their needs (Behr, 2019). Consider in this sense of a patient who can no longer care for themselves in the way they once could. Such a realization may trigger feelings of shame or humiliation and lead to emotional withdrawal from the caring relationship. It is precisely this defensive posture that humility can counteract. By helping the patient acknowledge their vulnerability, humility facilitates receptivity to care. Recognizing our fragility is a prerequisite for accepting help; likewise, understanding that we all exist along a fluctuating continuum of dependence, as Alasdair MacIntyre argues.

It is in this sense that Campbell and Swift define humility as the disposition of "acknowledging need for others" (Campbell & Swift, 2002, p. 32). For this reason, they include it among the "traits relating to the relationship between the self and others". According to the authors, it is a relational virtue insofar as it fosters the maintenance and building of relationships. This is also expressed by Brady, who asserts that: "[r]ealizing that one is vulnerable, dependent, and reliant upon others can lead to greater openness with people, gratitude for the attention and compassion that care givers show and strengthening of loving relationships" (Brady, 2018, p. 107). In addition, according to Carel, humility helps the ill person to understand the vulnerability and suffering of others (Carel, 2008, p. 33). For all these reasons, it has a powerful relational aspect too: it fosters the individual's openness to relationships, helps them recognize and appreciate the attention and care of others, and enables them to understand both others' suffering and their own limitations.

In addition, humility helps patients resist "otherwise excessive demands" (Kidd, 2012, p. 517) by cultivating what Mandelbaum calls "objectivity" (Mandelbaum, 1969, p. 162). Objectivity is crucial in illness, where false hopes can easily lead to harmful decisions. By fostering realistic self-assessment, humility enables patients to recognize when it is appropriate to halt burdensome treatments that no longer offer meaningful benefit.

Although this study works primarily on the virtues of the patient, it is important to emphasize that humility is equally relevant for healthcare professionals. Pellegrino (2007) makes this point forcefully: humility ensures that the "enormous powers of medicine to heal and harm" are not exploited for the physician's own interests, ego, or self-image. It helps the practitioner recognize the limits of medical science and serves as a counterweight to the arrogance that can compromise the dignity of the patient (Pellegrino, 2007, p. 76). Lebacqz echoes this point, noting that humility enables the physician to realize that "she needs the patient as much as the patient needs her [...] [t]he physician may know what the patient has, but only the patient knows what it means to have it" (Lebacqz, 1992, p. 302). This reciprocal recognition of vulnerability fosters more empathetic, collaborative, and ultimately more effective care. It also strengthens the trust that is essential to the therapeutic relationship[25].

Finally, humility is indispensable in guiding healthcare professionals to recognize when medicine can no longer offer a cure, when its role shifts to providing comfort and alleviating suffering at the end of life. In such moments, this trait allows clinicians to step back, accept their limits, and support the patient with dignity and compassion.

4. *Relational Virtues*

With the virtue of humility, my discussion of self-regarding virtues comes to an end. I will this now proceed to analyze the *relational virtues*. In this paragraph, I will focus primarily on the relational virtues that can enhance the patient-professionals relationship[26]. As emphasized in the first part of this work, the success of care depends on a variety of factors: the quality of the treatments provided, the

25 I have discussed the importance of testimonial justice in helping the physician to acknowledge reliability to the patient in Ricci, 2020.

26 For this reason, I have not included in the list other important relational traits such as forgiveness, which certainly play a significant role in the context of illness. This will be the subject of future works.

character of the patient, and on their openness to being cared for and to engaging in the caring relationship.

4.1. *Trust and Honesty*

As we have seen in the last paragraph, the virtue of humility plays a key role in opening up the subject, predisposing them to receive the caring gestures of healthcare professionals. However, it is not the only trait capable of doing this; *trust* and *honesty* are also crucial in this regard.

Trusting others and relying on their competence plays a fundamental role when the patient is dealing with a chronic or degenerative illness (see Navarini, 2020, p. 123). As we have seen in the first part of this work, caring is a relational work and is such precisely because it requires the commitment of both the carer and the cared-for in order to be truly effective. Even if the physician is competent, attentive, and morally grounded, their caring actions may ultimately be ineffective if the patient does not trust them or is unwilling to engage in the caring relationship.

It is worth noting that professionals can adopt a range of behaviors to foster trust, such as maintaining clear communication, being transparent with information, showing expertise, and demonstrating empathy. However, this is only part of the equation. The ill person must be willing to cultivate the dispositions necessary to enter into and maintain a trusting relationship. In this sense, trust is not merely an emotional state or a response to the behavior of others, but a trait of character that can be strengthened over time.

The presence (or absence) of trust can have significant repercussions on the patient's well-being and overall health outcomes. When patients trust their healthcare providers, they are more likely to adhere to treatment plans, communicate openly about their symptoms and concerns, and engage actively in the therapeutic process. Conversely, a lack of trust can lead to withdrawal, skepticism about medical advice, poor compliance with therapies, and ultimately, worse clinical outcomes. An interesting and illustrative example of this dynamic is provided by a study conducted by Andrea Benin and colleagues, which investigated the factors influencing mothers' decision to vaccinate their children. The research found that one of the key variables in the decision-making process was the level of trust mothers had in the physician. As the authors note: "[t]rust or lack of trust and relationships were main determinants of mothers' decisions about vaccination" (Benin et al., 2006, p. 1539).

This finding highlights that trust is not just a background sentiment but a central determinant in health-related choices. When translated to the clinical setting more broadly, it becomes clear that fostering and maintaining it is indispensable, not only for healthcare professionals, but also for patients, who must be disposed to enter into such a relationship and sustain it over time. In this sense, trust should not be thought of merely as reactive, but as an active moral disposition, a virtue that can and should be cultivated.

Another reason why is fundamental is its deep connection with the virtue of honesty. Although honesty appears in the literature under various labels (Jonathan Miles (2019) refers to it as 'veracity,' while Pellegrino and Thomasma (1993) prefer 'truthfulness') it broadly denotes the disposition to tell the truth, unless there may be reasons not to do so. Christian Miller offers a precise and influential definition of this trait: "Honesty is the virtue of being disposed, centrally and reliably, and as dictated by the capacities associated with practical wisdom, to not intentionally distort the facts as the agent sees them [...] along with the absence of significant non-virtuous motivation to distort the facts as the agent sees them" (Miller, 2021, p. 132).

Honesty plays a crucial role in the illness journey, particularly because it directly affects health outcomes. Indeed, a growing body of literature indicates that a lack of honesty can compromise the efficacy of treatments. Studies show that when patients distrust their healthcare providers, they are more likely to withhold or distort key information, such as their medical history, family background, existing conditions, or even lifestyle habits (see Lee & Lin, 2008, p. 44). This lack of transparency can lead to inappropriate or ineffective care, delayed diagnoses, and even medical errors. Being honest, then, is not only a matter of personal integrity, but it also becomes a collaborative act that sustains the therapeutic alliance. By disclosing truthful and accurate information, patients enable professionals to craft more appropriate and individualized care plans. On one side, when patients trust their physicians, they are more likely to share even sensitive or uncomfortable details. On the other, the more honest patients are, the better healthcare professionals will be positioned to understand and meet their actual needs, ultimately improving treatment outcomes and minimizing the risk of misdiagnosis (Battaglia et al., 2003; Hall, 2005).

Yet, the interplay between trust and honesty, though foundational, is only part of what makes trust such a pivotal virtue in the patient experience[27]. As Ying-Yang Lee and Julia Lin argue in their review on the role of trust in medical practice, trust serves as a catalyst for several mechanisms that empower patients to engage more effectively with the care process. One such mechanism is *compliance*[28].

Compliance, understood as the willingness to follow a proposed treatment plan, adopt necessary lifestyle changes, and face challenging procedures (e.g. chemotherapy, radiotherapy, surgery), is an indispensable part of the healing process. A trusting relationship increases the likelihood that the patient will embrace these measures, even when they are painful or demanding. Authors such as Pellegrino and Thomasma (1988) and Navarini (2020) highlight that trust allows the cared-for to fully benefit from the care offered. In this sense, a compliant patient will be more inclined to follow treatment protocols attentively, thus enhancing the effectiveness of the care they receive and better preparing them to respond to the challenges of illness.

The relationship between trust and compliance is fairly intuitive. Consider, for instance, a patient undergoing physiotherapy who, due to a perceived unreliability in the professional's behavior, struggles to establish trust. Even if the physiotherapist is highly

---

27 The relationship between trust and patient's health outcomes is shown by a large body of literature. In addition to the studies already cited see at least Birkhäuer et al., 2017; Lee & Lin, 2011.

28 The term "compliance" here refers to the patient's adherence to the treatment plan and dispositions proposed by the practitioner. See Lutfey & Wishner, 1999.

competent, with a robust track record and deep knowledge of pain management, this may not suffice. If the patient is inherently suspicious or unable to develop a sense of trust, they will likely be resistant to the proposed care plan. Conversely, when trust is established, the patient will be more inclined to follow their guidance and adhere to treatment[29]. Particularly interesting in this regard are clinical studies on diabetes mellitus. Lee and Lin elaborate on this:

> Focusing on patients with diabetes mellitus, who require significant alterations in lifestyle and completion of self-management tasks to achieve optimal control of their disease, patients with higher trust had better global assessments of overall ability to complete requisite diabetes care activities (Bonds et al., 2004). This explains why diabetic patients with higher trust in their physicians are more likely to better control their blood glucose (Alazri & Neal, 2003) (Lee & Lin, 2008, p. 44).

It should therefore be evident why the link between trust, compliance, and health outcomes is so crucial: trust encourages patients to follow treatment plans more faithfully, thereby enhancing their health outcomes.

Another important aspect that trust can provide is the professional's higher engagement in the caring relationship. As Lee and Lin have pointed out, this can play a significant role in determining the caring disposition of professionals. In fact, being thought of as a trustworthy person can increase the responsibility physicians perceive towards patients, their commitment, and, as a consequence, the effectiveness of their performance. This last aspect turns out to be particularly interesting because it shows that illfullness is beneficial not only for the patient, as it can also foster the practitioners' flourishing. In fact, although physicians should be well-disposed and caring toward their patients, not all practitioners are moral exemplars. Therefore, although the importance of moral education in clinical settings cannot be

29 On the relationship between trust and compliance see e.g. Safran et al., 1998; Trachtenberg et al., 2005.

overstated, the patient's character and attitude toward the care relationship remain essential to enhancing its effectiveness.

From this point of view, Lee and Lin's research also shows that care really is a relational practice in that the contribution of both the carer and the cared-for is crucial.

Finally, according to the authors, patients can also be positively influenced (again through the mediation of the virtue of trust) by what they refer to as the "placebo effect" (see Lee & Lin, 2008, p. 44). As literature shows, the greater the trust in professionals, the greater the benefit patients reap from treatments happens in that "[b]esides the real effects of medication or surgery on the patient, there may be an additional placebo response which will further improve the inherent capacity for self-healing" (Lee & Lin, 2008, p. 44). Thus, patient-professionals relationship acts almost like a medicine. Trust can lead the patient to internalize the belief that they are improving, which in turn enhances their perceived well-being and can trigger genuine physiological benefits.

Given these reflections on trust and its centrality, not only for the therapeutic alliance but for health outcomes more broadly, I present below the figure that Lee and Lin use to summarize the various mechanisms through which trust in healthcare professionals influences patients' well-being.

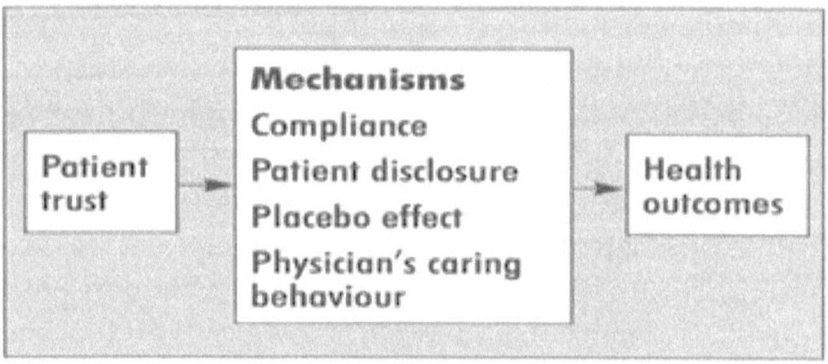

Fig. 1 Lee & Lin, 2008, p. 44

As we have seen, trust and honesty are essential traits in navigating life in illness. However, it is important to reiterate that these virtues must be guided by reason to avoid slipping into excess or deficiency. An excess of trust, for instance, may lead the patient to lose sight of their own needs and preferences, failing to assess whether a proposed treatment aligns with their values or whether they have the emotional and physical resources to pursue it. In some cases, excessive trust may even result in overtreatment or in unrealistic expectations placed on the healthcare professional, potentially fostering paternalistic dynamics. For this reason, while patients should indeed trust healthcare professionals, it is equally crucial that professionals avoid overriding patients' decision-making capacities. Trust should not silence judgment. Instead, it should coexist with the patient's reflective engagement. As Navarini notes, "trust is a virtue when the subject-patient acquires a good capacity to understand how much to believe, that is, to what extent to confide in the carer" (Navarini, 2020, p. 123, my translation). After all, it is precisely the ability to strike the right balance between extremes that characterizes virtuous traits.

4.2. *Gratitude*

Having described the virtues of trust and honesty, I now turn to the final virtue addressed in this chapter: *gratitude*[30]. Despite its importance, gratitude remains a markedly underexplored and often neglected virtue in the context of illness[31].

It has been defined in the psychological literature as "a personality disposition to notice, appreciate and respond with grateful emotions and thankfulness to other people's acts of benevolence and kindness" (Ruini & Vescovelli, 2013, p. 264)[32]. However, gratitude also emerges more broadly as an expression of the subject's appreciation of life itself (Kim et al., 2016). As Brady notes, this virtue can only manifest when the individual has reached a sufficient degree

---

30 Gratitude has also been understood as an emotion, for example, by Kristjánsson (2018) who understands it specifically as a "virtuous emotion".

31 This appeared more extensively in Navarini, 2020, but only as hinted at in other authors such as Carel, Kidd, and Miles.

32 The authors drew this definition from the various elements that emerged from researches by McCullough and colleagues (2002).

of openness to receiving care: "Realizing that one is vulnerable, dependent, and reliant upon others can lead to greater openness with people, gratitude for the attention and compassion that caregivers show, and strengthening of loving relationships" (Brady, 2018, p. 107). In this sense, both humility and trust are foundational: humility fosters the receptivity necessary to welcome the "gift" of care, while trust promotes the cared-for's willingness to engage in a genuine relationship with the carer. The more the patient trusts the professional, the more attuned they become to recognizing the care they receive, and the more likely they are to respond with gratitude. Trust, in other words, can activate the very mechanism that gives rise to this virtue.

We can thus say that gratitude has a generative and reciprocal quality. It is capable of setting in motion a self-reinforcing virtuous cycle: the more grateful the patient is, the more likely the caregiver is to respond with further acts of benevolence. As Liz Gulliford puts it, "gratitude helps build social bonds and makes people feel valued" (Gulliford, 2020, p. 101). Furthermore, a substantial body of psychological research confirms its relevance not only in the context of caregiving relationships but also as a critical factor in coping strategies and psychosocial recovery following trauma. For instance, a study by Chiara Ruini and Francesca Vescovelli demonstrates that gratitude can support breast cancer patients in achieving greater well-being by fostering post-traumatic growth (PTG). This process enables individuals to derive new meaning from their experience and to re-engage with life in a purposeful way[33]. Gratitude has also been shown to correlate positively with resilience and character development, and negatively with depression (e.g. Fredrickson et al., 2003).

This chapter has aimed to identify and explore the virtues most essential to life in illness, those that may enable individuals to navigate the challenges it poses with strength, clarity, and moral purpose. Below, I include a diagram that summarizes the two clusters discussed.

33 On this point see also Emmons & Crumpler, 2000.

It is possible that some readers may find this list incomplete, feeling compelled to add or remove certain traits. However, this should not be seen as a limitation. No list of virtues can ever be final, especially in a moral domain that is deeply shaped by cultural, temporal, and societal contexts. In the next and final chapter, I will explain why this lack of definitive closure is not a problem for the framework presented here.

Self-regarding virtues	Aims
Hope	Aids in finding a sense of meaning and purpose in life during illness, as well as in identifying new goals to pursue and pathways to achieve them, even when this requires changing direction multiple times.
Courage	Compels to take risks when faced with complex decisions or actions that involve significant dangers.
Perseverance	Equips the ill individual with the endurance necessary to achieve goals that may be long-term and challenging to accomplish.
Humor	Helps reframe negative experiences, regulate difficult emotions, and maintain a future-oriented perspective. It facilitates emotional resilience, promotes adaptive coping, and fosters positive interactions with others, thereby supporting psychological well-being throughout the illness experience.
Patience	Regulates one's emotions. It also helps the ill subject to handle the waits and to cope with the various expectations that arise during the illness journey, without succumbing to anxiety or frustration. Additionally, it strengthens the relationships between the individual and others by preventing the former from projecting their frustrations onto those around them.
Humility	Involves accepting one's limitations, vulnerability, and dependency. It is particularly significant because it fosters a willingness to be open to the care and support of others.

Tab. 3

Relational virtues	Aims
Trust	Facilitates open dialogue with the practitioner, promoting compliance and enhancing the relationship with healthcare providers, which in turn helps deliver the best possible care. Additionally, trust has a placebo effect, leading patients to perceive higher levels of well-being, and encouraging them to activate the virtue of honesty.
Honesty	Allows patients to provide accurate information to healthcare professionals, facilitating the formulation of correct diagnoses and ensuring that the cared-for receive the best possible care.
Gratitude	Is one of the most powerful engines for caring actions, as it can create a virtuous circle. When the carer senses gratitude, they tend to be more generous and become increasingly involved in caring for both the patient and the patient's condition, which, in turn, fosters even greater gratitude in the cared-for.

Tab. 4

CHAPTER 6
A POSSIBLE SOLUTION FOR THE CHARACTER EDUCATION ON THE ILL PERSON

The list of virtues proposed here brings with it issues that afflict all such lists. Although the effort of extrapolating a set of traits to cope with the experience of illness and foster positive coping is essential, the identification of them may come across as non-satisfactory. Whenever a list of traits is proposed, there is always an underlying dissatisfaction as the list is likely to be incomplete. It is difficult, arbitrary, and even somewhat presumptuous to imagine that one establishes a definitive list of traits that applies to everyone in all contexts. It will always lead to the question of how it can be modified, whether it can benefit from a reduction or whether an extension is appropriate. Unlike other domains, such as biology, where one might aim for a comprehensive enumeration of anatomical structures, the moral sphere resists this kind of closure. Morality does not lend itself to exhaustive categorization because the terrain it encompasses is far more expansive and fluid. In this sense, completeness in moral theorizing remains an aspirational ideal rather than an achievable endpoint.

This is not to suggest that the attempt to outline a set of relevant traits is without merit. As previously discussed, the proposed lists offer a useful indication of those dispositions that may be particularly effective in supporting individuals as they navigate the challenges of life in illness. However, what ultimately proves most significant is not the specific enumeration of virtuous traits, which will inevitably vary across contexts, but rather the common denominator that underlies them, a quality more fundamental than the traits themselves. Regardless of which virtues are included or omitted, what remains essential is the element of *virtuosity*, the underlying moral competence that enables agents to manage and respond appropriately to the demands of illness. What is crucial, then, is not the pursuit of an exhaustive or definitive catalogue of virtues, but the cultivation of the capacity to discern how to act rightly across a wide range of circumstances. This meta-level capacity manifests as a set of interconnected abilities: *emotional regulation*, *creative adaptability*, *situational discernment*, and the *aptitude to recognize which dispositions are most relevant in a given context*.

Virtuosity is constituted by the ability to integrate and apply moral insight across varying conditions and can be identified in *phronesis,* or practical wisdom.

Phronesis is thus the common element that it is possible to find in every single trait I have mentioned, and it can be found in all virtues because it is their *raison d'être*, or what we may Kantianly call their *ratio essendi* (De Caro et al., 2024). Indeed, practical wisdom constitutes the deepest nucleus of each virtuous trait. For this reason, we could certainly say that virtuosity consists essentially in *phronesis*, which manifests itself in various circumstances through the virtues (De Caro et al., 2025). Accordingly, the virtues may be understood as the modes in which *phronesis* becomes operative in context-sensitive judgment and action.

1. *An Overview of the Aretai Model*

This suggestion that virtuosity would consists of practical wisdom understood as ethical expertise has been proposed by the Aretai model (De Caro et al., 2018; 2021; 2024; 2025; De Caro & Vaccarezza, 2020). The Aretai model defends "the conceptual, epistemological and ontological priority of practical wisdom" (De Caro et al., 2021, p. 29) and thus "pivots on three main tenets: (i) Virtue monism: ontologically, *phronesis* is the only moral virtue; (ii) Virtue molecularism: *phronesis* manifests itself through clusters of individual moral virtues; (iii) *Phronesis* is a particular kind of expertise" (De Caro et al., 2025).

In this sense the Aretai model characterized as distinct from "The Standard Aristotelian View" (SAV), which conceives virtues as distinct from *phronesis.* According to this view, the virtues benefit from *phronesis*, which represents "the condition of moral-cognitive excellence" (see De Caro et al., 2018, p. 291). Virtues require *phronesis*, yet *phronesis* itself is not sufficient without the virtues. According to the SAV, the virtues are thus indispensable because they guide practical wisdom in discerning and pursuing the good across the various domains, contexts, and situations of life. By contrast, according to the "Aretai model" what ultimately defines the virtuous moral agent is not the possession of a set of discrete virtues, but rather the unified capacity of ethical expertise. Ad De Caro and colleague write: "The only trait a virtuous moral agent really possesses is practical wisdom (*phronesis*) as ethical expertise – a trait that is manifested at various degrees in different domains and whose description can be given in terms of clusters of traditional virtues, though its ultimate nature is unitary (De Caro et al., 2021, p. 31).

This means that the only virtue is practical wisdom (conceived as ethical expertise). The other so-called "virtues" are, in this light, context-specific designations, i.e. terms we use to indicate how *phronesis* manifests and adapts across the diverse domains and situations of moral life.

According to this view, the virtuous individual (the *phronimos*) is an ethical expert who possesses a unified practical skill that enables them to intuitively grasp how to act across the full spectrum of moral experience. A defining feature of such moral expertise, as already noted, is the ability to discern which particular trait (or aspect of practical wisdom) is called for in a given situation, without the need for deliberate cognitive effort, unless the situation is particularly complex, as in the case of moral dilemmas (De Caro et al., 2021).

This capacity to see what is morally required, without having to weigh multiple alternatives, has been described by John McDowell as the "silencing power of virtue" (McDowell, 1998, p. 17). To make as example, when a virtuous agent is confronted with the opportunity to return a lost wallet, they do not consider "running off with it" as a viable option. That possibility is cognitively muted: they immediately see that returning the wallet is the right course of action. The expert, in this sense, sees the moral landscape clearly and responds accordingly, without hesitation.

Yet, as De Caro and colleagues emphasize, this immediate responsiveness may not hold in situations that are unfamiliar or particularly challenging. In such cases, especially when the agent is faced with a moral dilemma or an unprecedented context, the subject may require deliberation or may even find themselves temporarily disoriented. Moral expertise does not entail infallibility: the virtuous person, like anyone else, may lack the resources to choose what to do in wholly novel circumstances. Thus, while it is essential to recognize a threshold of moral competence beneath which a subject can no longer be considered an expert, it is equally important to acknowledge that ethical expertise is dynamic and exposed to growth and refinement over time. As the authors note, the criterion that renders a subject virtuous is not moral perfection, but rather the ability to recognize one's limitations and to strive to overcome them. De Caro and colleagues have defined this capacity as "openness to new domains and situations" (De Caro et al., 2021, p. 31).

Hence, we can conclude that they conceive virtuosity as an ethical competence (i.e. a form of moral expertise) that enables the subject to understand how to act in the different contexts that pertain to morality. Like other form of expertise, it can be improved, since becoming an expert implies a progressive competence that gradually extends to new domains.

2. *How Virtue Monism Can Rescue Virtue Theory from Its Critics*

The Aretai model is interesting not only because it fosters a realistic understanding of virtuosity, it also responds to a number of problems that have challenged more traditional account in virtue ethics. Virtue Monism, in particular, seems to be able to answer three challenges (De Caro et al., 2021) namely the *situationist challenge*, the *anti-rationality challenge* and the *automaticity challenge*[1].

First of all, the Aretai model is able to respond to the critique posed to virtue ethics by the situationists. As De Caro and colleagues note, traditional virtue theories often conceive of virtues as distinct and ontologically separate traits, each governing a particular domain.

1 On this point, see also Niccoli, Piantoni & Ricci, 2024.

According to virtue atomism, these traits should reliably guide the subject's moral responses across a wide range of contexts:

> Whether one is the captain of a sinking ship, the witness of an abuse of power, or a shy student who just graduated from high school and who must draw on all her strength to pack up and move away from home to cultivate her talent at a faraway college, all acts of courage are identical according to the virtue atomist, and the very same virtue – if truly acquired – has to be reliably manifested in all these situations (De Caro et al., 2021, p. 30).

Situationists challenge this assumption, arguing that empirical evidence suggests character traits are highly context-dependent. Because behavior is heavily shaped by situational factors, what appears to be a stable trait may in fact be a narrow disposition tied to specific circumstances. If this were true, it would necessitate positing an unmanageable proliferation of virtues (such as courage-in-battle, courage-in-illness, courage-in-fire) and would severely strain the plausibility of traditional models.

The Aretai model responds by rejecting the atomistic view of character. Instead of imagining character as composed of isolated traits, it understands virtue as an expression of unified ethical expertise that manifests flexibly across diverse domains. Rather than demanding moral perfection or consistency across every conceivable situation, the model posits that a virtuous person must possess a sufficient threshold of competence along with an openness to continued moral development. As De Caro and colleagues argue, any model requiring complete consistency would result in an implausibly idealized agent, an "implausible strawman" (Upton, 2009, p. 109). In this sense, the Aretai model not only withstands the situationist critique but also offers a more nuanced and empirically plausible alternative.

The second challenge is the anti-rationality objection. Since the Aretai model places central importance on practical wisdom, one might worry that it reverts to a hierarchical model of the mind, in which reason is elevated above other cognitive or emotional capacities.

De Caro and colleague respond to this critique by clarifying that, although virtue monism undoubtedly grants significant importance to reason, this does not entail a denial of the relevance of other mental states; rather, it rejects any hierarchical view of the mind. The emphasis on practical wisdom should not be construed as an appeal to rationalism. On the contrary, emotions are, according to De Caro and colleagues, essential "both in the process of formation of moral character (De Caro et al., 2018) and in the attunement with an agent's concrete situation that makes it possible for her to see what moral options are in play" (De Caro et al., 2021, p. 40). From this perspective, several components are involved in virtuous action: not only different kinds of knowledge but also the "affective orientation to virtuous ends, as well as a cognitive-emotional ability to perceive the moral requirements at stake in a given situation" (De Caro et al., 2021, p. 40). Thus, the Aretai model is not vulnerable to the anti-rationality challenge.

A final challenge addressed by this theoretical framework has recently been raised in the context of virtue theories, namely the so-called *automaticity challenge*. As discussed in chapter five of this work, Stichter offers an insightful response to this problem. In similar fashion, De Caro and colleagues clarify that automaticity in moral decision-making is not to be understood as a detachment from rational processes, but rather as the result of a high level of moral expertise gradually acquired through practice. In this sense, consider someone aspiring to become an expert musician. They may start practicing by mechanically repeating a series of gestures. However, as their knowledge improves and these gestures become assimilated, they no longer need to consciously focus on them. A similar process occurs with the moral expert, who gradually internalizes, through practice, the right way to respond morally in different situations[2]. Initially, they will need to be guided by rules, principles, or imitation

---

2 Although the monist account does not explicitly emphasize the role of feedback in moral practice, I would argue that it can play a crucial role in refining one's ethical expertise. Consider, for instance, a professional athlete preparing for a competition: feedback from coaches and peers becomes an indispensable resource. Precisely because the athlete is already well-trained and accustomed to performing certain movements, the risk of reinforcing subtle but persistent errors is higher than for a novice. Similarly, even highly virtuous individuals may benefit from the

of moral exemplars. For instance, a child may explain their helping behavior by stating: "I helped that person because mom says I have to be kind to my friends". However, through continued practice and engagement with morally salient situations, individuals progressively acquire higher levels of ethical expertise. Over time, they develop an internalized moral competence that allows them to respond directly and appropriately to the demands of a given context. As De Caro and colleagues observe, "the virtuous person develops a sensitivity to moral reasons such that she needs to weigh them with deliberation only in situations that are particularly challenging or uncharted" (De Caro et al., 2021, p. 42). The Aretai model thus does not endorse a notion of intuition detached from reason. Rather, intuition is seen as the refined outcome of a process of rational and emotional attunement. Consequently, the so-called automaticity challenge does not undermine the theoretical consistency of the model.

3. *Moral Experts and Life in Illness*

In the previous paragraph, I clarified the main tenets of the Aretai model and outlined the reasons why this framework offers a particularly compelling account. Not only does this model provide a more concrete and realistic vision of the kind of moral excellence to which human beings can aspire, but it also proves resilient in the face of some of the most pressing objections raised against SAV. Having laid this foundation, the next step is to explore how this theory can help to account for the patient's illfullness.

As previously discussed, virtuosity does not reside in the mere possession or display of discrete character traits, but rather in the wise and context-sensitive management of one's moral life. As De Caro and colleagues write, "when we label someone as, say, courageous, what we really admire is one's being wise (i.e. ethically expert) in the field of danger and fear, broadly conceived; courage is the name of wisdom in that particular domain" (De Caro et al., 2018,

---

insights of other morally expert agents, who can help identify blind spots and promote further growth in ethical competence.

p. 294). The reference to a given virtue thus serves an explanatory and classificatory function: it allows us to describe and communicate the nature of the agent's conduct within a specific moral context.

With regard to the experience of illness, it may be argued that what is most fundamentally required for the subject to cope effectively, and navigate its complexities is practical wisdom understood as ethical expertise. It is *phronesis*, in this sense, that most aptly equips the patient to respond creatively and intelligently to the new circumstances that life imposes on them.

This raises an important question concerning the status of the virtues themselves: if practical wisdom is understood as the core of ethical expertise, what role (if any) remains for the individual "virtues" themselves?

As previously argued, illfullness can be fostered through the cultivation of practical wisdom, which enables the subject to respond appropriately to the moral and emotional demands of illness. *Phronesis* equips patients to navigate complex and evolving circumstances, to find adaptive and creative responses even in the face of repeated setbacks and relapses, and to deliberate effectively in morally dilemmatic situations. Furthermore, it underpins the capacity to regulate one's emotions, an ability particularly salient in the context of suffering and vulnerability.

Nonetheless, I argue that this should not entail that virtuous traits should be dismissed or regarded as superfluous. On the contrary, it is important to recognize that the acquisition of practical wisdom requires time, experience, and sustained moral development. In cases of serious or rapidly progressing illnesses, however, time may be a scarce resource. Under such conditions, virtue-oriented interventions can still play a valuable role in supporting the patient's ethical and psychological adaptation. In this sense, identifying those traits most relevant to the specific vulnerabilities or emotional challenges introduced by the illness[3] can inform targeted character education programs. For example, ass discussed, hopelessness has

3 For example, it has been shown that depression and anxiety have pathophysiological links with Parkinson disease. On this see e.g. Martínez-Martín & Damián, 2010.

been shown to correlate strongly with depression, suicidal ideation, and increased psychological distress (see Beck, 1986; Beck et al., 1975). In such cases, fostering hope through structured interventions may not only improve the patient's emotional well-being but also reduce the risk of more severe psychological outcomes.

What follows is that ideally, the cultivation of practical wisdom as a form of ethical expertise should begin at the earliest possible stage, namely from the moment of diagnosis. Early character education interventions aimed at fostering *phronesis* would provide the ill person with the resources necessary to navigate the moral and existential dimensions of illness. Nevertheless, not all patients are afforded the temporal or psychological conditions necessary for such sustained development. For those facing a limited prognosis, alternative pathways must be considered in order to focus more narrowly on the cultivation of specific traits that can support their well-being and foster a sense of self-efficacy. This approach does not entail a return to rigid lists of virtues, but rather acknowledges the significance that certain traits may hold within the specific context of illness. As we have seen, even virtuous individuals may nonetheless exhibit areas of vulnerability. In such cases, targeted work on particular traits may serve as a valuable complement to their broader ethical competence.

This approach, while contextually necessary, may have its limitations. The cultivation of isolated traits in a short timeframe may not suffice to develop *phronesis* in its full expression. Yet, this does not diminish its relevance. When time is short, the ethical imperative may no longer be to form a comprehensive moral expert but to enable the subject to live meaningfully and with dignity. The cultivation of specific virtuous traits may thus play a crucial role in supporting the patient's ability navigate adversity.

Before concluding this chapter, it is important to reiterate a point previously underscored in this work: the cultivation of virtues must take place within appropriate contexts that address not only the patient's physical suffering, but also the broader existential, psychological, and moral dimensions in which distress may emerge. A particularly well-suited setting in this regard is early palliative care. Indeed, as previously underlined, early palliative care offers the potential to accompany patients through a trajectory that promotes holistic well-being. To realize this potential, however, it is essential to foster a cultural shift in how palliative care is perceived. Rather than being seen merely as a site of terminal decline (e.g. Fliedner et al., 2021; Sanjo et al., 2008; Shalev et al., 2018; Zimmermann et al., 2016), palliative care should be recognized as a space where patients can cultivate a meaningful life within illness. As Navarini emphasizes, we must "increasingly distance ourselves from the provision of pain therapy alone and increase the tools to promote moral, spiritual and psychological well-being" (Navarini, 2020, p. 121, my translation). Early palliative care can thus support patients in taking care of their inner world, potentially from the moment the diagnosis is communicated. This may allow them to experience better management of their condition and consequently to achieve higher levels of well-being. Palliative care can thus be the appropriate venue for a virtue care ethics program.

Importantly, this should not imply that character work is to be imposed. On the contrary, this is antithetical to the development of genuine moral dispositions. Rather, what is needed is the institutional provision of real opportunities, spaces and practices that allow each patient to take care of their flourishing. This vision is already implicit in the foundational philosophy of Cicely Saunders, who emphasized the importance of fostering the possibility of flourishing until the very end.

CONCLUSIONS

The aim of this work has been to explore the compatibility between illness, suffering, and moral flourishing. To this end, I began by identifying a theoretical framework capable of accommodating such a perspective. I thus introduced virtue care ethics; this integrated approach allows to preserve the relational and contextual insights of care ethics, with virtue theory's focus on character development. As shown, care ethics constitutes a vital framework, as it foregrounds the centrality of care in human life and reminds us of our fundamental interdependence. Furthermore, it emphasizes that care is, at its core, a relational practice whose effectiveness depends not only on the caregiver's competence but also on the care receiver's openness to engage in the caring relationship. In this sense, character formation proves essential for both the caregiver and the cared-for.

Nevertheless, despite its merits, care ethics alone does not suffice. One of its main limitations lies in its lack of attention to character education. Indeed, care ethicists assume that human beings are capable of caring gestures simply because they have been cared for. However, this does not account for the reflective, effortful, and often challenging work required to develop into a caring agent. For this reason, I have argued that for this theory to work in caring context, it must be complemented by other theories. Virtue ethics has proven to be as particularly suitable in this respect, providing normativity and systematicity.

Virtue ethics, thanks to its focus on character, offers a normative framework that is particularly well-suited to contexts in which the development of moral dispositions is essential. This theoretical perspective not only identifies which traits are most conducive to flourishing within specific domains but also provides guidance on how to cultivate them. In this sense virtue care ethics proves particularly effective in caring contexts as it preserves care ethics' attentiveness to the centrality of care as well as its relational and contextual approach, while recognizing that good care cannot occur without the cultivation of character.

In the second part of this work, I turned to the core of the relationship between illness and virtue, beginning by distinguishing the experience of suffering from that of pain. As emphasized, although these two terms are often used interchangeably, they refer to distinct dimensions of human experience: pain is primarily associated with the biophysical domain, whereas suffering operates at a deeper existential level and is marked by a persistent and intense sense of distress. One of the defining characteristics of suffering is precisely its pervasiveness; this feature – that led Saunders to describe it using the expression total pain – can produce a kind of existential "short circuit" in the ill subject, evoking a sense of being trapped in a condition perceived as inescapable, thereby compromising both well-being and the possibility of happiness. However, as suggested, this is not the only possible trajectory. While illness is indeed a pervasive condition, it does not necessarily extinguish the individual's capacity to flourish and lead a meaningful life.

This is not to suggest that the experience of illness should be trivialized. Coping, as is well known, may be either adaptive or maladaptive, that is, it can manifest as a positive or negative response to the circumstances imposed by illness. For this reason, it is essential to adequately support patients throughout their care journey. With appropriate support, not only can the subject engage in positive adaptation, but they may also experience PTG. PTG has been shown to enhance perceived well-being and foster a meaningful transformation in the patient's character through the development (or consolidation) of specific traits. Illness, in this sense, may harbor unexplored opportunities for flourishing. As several scholars have noted, the encounter with illness and suffering can make individuals more receptive to the cultivation of certain virtues, such as perseverance. This is due to the fact that this experience often entails both minor and major daily challenges which, over time, can shape and reshape the individual's personal and narrative identity. From this perspective, suffering may not merely be an impediment to flourishing but, under certain conditions, it may become a pathway to virtuosity.

The final part of this work focused on identifying a list of character traits particularly relevant to this context. In doing so, I intended to show that virtues are not only the outcome of a positive adaptation to the experience of illness but can also function as protective factors helping individuals manage the challenges of their condition more effectively. Indeed, I emphasized that each domain of life constitutes a distinct sphere of moral experience and, as such, calls for the development of specific traits, or, more precisely, of particular shades of practical wisdom that are responsive to the demands of that domain. In this sense, the experience of illness similarly calls for the cultivation of character traits that enable the subject to endure, make sense of, and navigate the transformations imposed by it.

This account proves especially attentive to the lived reality of moral life. It also does not presuppose that the agent possesses full moral competence across all domains to be virtuous. On the contrary, what is required is the capacity to recognize one's limitations and a corresponding openness to moral growth. Certain domains, such as illness, can only be fully understood through lived experience. It is precisely for this reason that illness has been described in the literature as a "transformative experience", one that can radically reshape personal and narrative identity. The transformative and demanding nature of illness may render the individual unprepared to confront its challenges. Accordingly, the subject may require time to adjust to the new condition and to cultivate the emotional and moral capacities necessary to navigate it.

I finally proposed a list of traits that can help the ill subject to better manage illness and its consequences. I identified these traits in two cores: *self-regarding virtues* and *relational virtues*. In the first cluster, I included hope, courage, perseverance, humor, patience, and humility.

Among the self-regarding virtues, hope emerges as a foundational trait. It enables the patient to reorient their perspective, allowing them to maintain a sense of meaning and purpose. Hope is also a crucial "virtue" as it allows the subject to identify new goals and to discern viable pathways toward their realization, even when doing so requires revising expectations or shifting trajectory multiple times. Importantly, hope fosters the capacity to rise after setbacks, thereby fostering resilience.

However, given that the illness trajectory is often punctuated by complex moments, emotionally charged decisions, and significant fears, hope alone may not suffice. It must be sustained and complemented by other traits that can assist the subject in translating intention into action. For this reason, courage and perseverance are equally indispensable. Courage equips the patient to "step into" difficult and sometimes frightening situations, such as undertaking invasive treatments or consenting to emotionally and physically demanding therapies. Perseverance, on the other hand, is what enables the subject to continue along these paths even when progress is slow or motivation falters. Let us consider a patient who has relapsed: they would not undertake a new treatment if they had no hope in disease remission, symptoms control, or if they were not convinced that treatments would allow them to have more time to spend with their children and family. The fundamental importance of hope is evident in this example. Nevertheless, when a new path (such as a new therapy) is proposed to them, they may be frightened by its heavy consequences, such as losing their hair or no longer recognizing themselves in their new body. It is precisely for this reason that the virtue of courage is fundamental. Furthermore, achieving desired goals takes time and requires the ability to endure setbacks and moments of lost motivation. This calls for the virtue of perseverance.

I also underscored the relevance of humor, particularly as a resource for emotional regulation and for the re-signification of difficult experiences. When appropriately cultivated, humor can help the patient to confront suffering without being overwhelmed by it, offering a different perspective from which to interpret their condition. Humor also facilitates access to complex emotions, fosters resilience, and supports the construction of a livable narrative even in the midst of adversity.

Finally, I included among the self-regarding virtues the traits of patience and humility. Patience emerges as an essential disposition, particularly due to its capacity to regulate emotions. This regulatory function helps the patient to face the illness trajectory without being overwhelmed by anxiety, frustration, or discouragement. Moreover, by tempering impulsive emotional reactions, patience may prevent the kind of interpersonal tensions that might arise from the strain of chronic or acute suffering and could compromise the patient's relationships with caregivers, loved ones, or healthcare professionals. In this sense, patience functions as a bridge between self-regarding and relational virtues, as it contributes not only to the inner stability of the subject but also to the preservation and cultivation of meaningful connections with others.

Humility, understood as the capacity to accept one's own limitations, vulnerability, and dependency, plays a crucial role in illness, as it fosters the subject's openness to receiving gestures of care. It enables the patient to acknowledge the need for support and to accept the condition of being cared-for without perceiving it as a threat to their dignity or autonomy.

Turning to the relational virtues, I included within this cluster the traits of trust, honesty, and gratitude. As previously discussed, care is fundamentally a relational practice: its effectiveness depends not only on the caregiver's disposition and technical competence but also on the cared-for's character as well as on their openness and willingness to engage in caring relationship.

Among the relational virtues, trust occupies a central place, as it has been shown to promote positive health outcomes and enhance the patient's overall well-being. Trust enables patients to communicate openly with practitioners, creating the conditions for more accurate diagnoses and more tailored treatments. Furthermore, it predisposes the patient to activate the virtue of honesty, another fundamental trait in the context of clinical care. Honesty, by fostering truthful and comprehensive communication, helps ensure that caregivers receive all the relevant information needed to develop an appropriate and effective care plan.

The final relational virtue I included is gratitude. As observed by Gulliford (2020), gratitude is one of the most powerful motivators of caring behaviors as it often gives rise to a virtuous circle: when caregivers perceive genuine gratitude from the patient, they are more inclined to act generously and to engage more fully in the patient's care. This in turn elicits deeper gratitude in the patient, reinforcing the caring dynamic and strengthening the bond between carer and cared-for.

I finally concluded the third part drawing attention to some limitations that inevitably affect any attempt to construct lists of virtues. While identifying a set of traits that may assist the patient in navigating illness is undoubtedly valuable, such lists cannot be considered definitive. However, this does not represent a real problem. What ultimately proves most significant is not the specific enumeration of virtues, but rather the underlying quality they share, virtuosity (which corresponds to practical wisdom). As I have argued, practical wisdom enables the subject to regulate emotions, to discern how to act appropriately across different moral contexts, and to understand what traits may be most relevant in a given situation.

While the most desirable path may consist in cultivating practical wisdom in this context, this route is not always accessible. The development of *phronesis*, even of its context-specific expressions, requires time, reflection, and sustained moral engagement. These, however, are resources that may be limited or altogether unavailable in the context of serious or rapidly progressing illness. Furthermore, as previously noted, even individuals who possess a certain degree of practical wisdom may nonetheless exhibit specific areas of moral fragility. In such cases, a targeted effort focused on individual traits can still prove meaningful, offering concrete support in navigating the moral and emotional demands of illness. In this sense, it becomes essential to understand what kind of character work is most appropriate for which patient.

In conclusion, this work did not intend to offer a definitive account of the themes addressed, rather, it aimed to provide a conceptual foundation for future investigations. The reflections presented in this work constitute only the preliminary steps of a wider philosophical and empirical investigation. While some possible directions have already been outlined throughout the text, the central aim here has been to provide the conceptual foundations for a virtue care ethics framework which is, by definition, open to empirical work and thus to the creation of protocols and programs helping patients work on their character and experience illfullness.

Although some interventions focusing on the development of individual traits have been developed – such as Charles Snyder's Hope Therapy and Max Chochinov's Dignity Therapy – it would be valuable to design an intervention protocol that adopts a mixed approach and aims to cultivate a broader set of traits. The goal would be to support the patient's moral expertise and strengthen their capacity to navigate the experience of illness. One part of the protocol could aim to foster the recognition of the centrality of care and support the development of practical wisdom. This would include focused work on moral identity, motivation, emotional regulation, and moral sensitivity. Furthermore, a specific module could be devoted to the cultivation of individual traits (whether self-regarding or relational) that are particularly relevant to the illness experience, enabling patients to address specific areas of vulnerability, even when time is limited. Developing such a protocol could represent a meaningful direction for future research in virtue care ethics.

REFERENCES

Adams, R. M. (2009). *A Theory of Virtue: Excellence in Being for the Good*. Oxford University Press.

Alfano, M., Astola, M., & Urbanowicz, P. (2022). Having a Sense of Humor as a Virtue. *The Journal of Value Inquiry, 58*, 659-680. https://doi.org/10.1007/s10790-022-09918-1

Allen, C. (2004). Animal Pain. *Noûs, 38*(4), 617-643. https://doi.org/10.1111/j.0029-4624.2004.00486.x

Allmark, P. (1998). Is Caring a Virtue?. *Journal of Advanced Nursing, 28*(3), 466-472. https://doi.org/10.1046/j.1365-2648.1998.00803.x

Annas, J. (1995). Virtue as a Skill. *International Journal of Philosophical Studies*, *3*(2), 227-243. https://doi.org/10.1080/09672559508570812

Annas, J. (2011). *Intelligent Virtue*. Oxford University Press.

Beermann, U., & Ruch, W. (2009). How virtuous is Humor? Evidence from Everyday Behavior. *Humor, 22*(4), 395-417. https://doi.org/10.1515/HUMR.2009.023

Bain, D., Brady, M., & Corns, J. (Eds.). (2019). *Philosophy of Suffering: Metaphysics, Value, and Normativity*. Routledge.

Bates, V. (2018). 'Humanizing' Healthcare Environments: Architecture, Art and Design in Modern Hospitals. *Design for Health*, *2*(1), 5-19. https://doi.org/10.1080/24735132.2018.1436304

Beauchamp, T. L., & Childress, J. F. (2001) *Principles of biomedical ethics*. Oxford University Press.

Beck, A. T. (1986). Hopelessness as a Predictor of Eventual Suicide. *Annals of the New York Academy of sciences*, *487*(1), 90-96. https://doi.org/10.1111/j.1749-6632.1986.tb27888.x

Beck, A. T., Kovacs, M., & Weissman, A. (1975). Hopelessness and Suicidal Behavior: An Overview. *Jama*, *234*(11), 1146-1149. https://doi.org/10.1001/jama.1975.03260240050026

Benin, A. L., Wisler-Scher, D. J., Colson, E., Shapiro, E. D., & Holmboe, E. S. (2006). Qualitative Analysis of Mothers' Decision-Making About

Vaccines for Infants: the Importance of Trust. *Pediatrics*, *117*(5), 1532-1541. https://doi.org/10.1542/peds.2005-1728

Benner, P. (1994). From Novice to Expert. In M. Brown, F. Patrick, D. Tate, & S. Wright (Eds.). *A Collection of Readings Related to Competency-based Training* (pp. 129-137). Deakin University.

Bina, F., Bonicalzi, S., Loria, M., Palazzolo, C., & Ricci, E. (2024). La nozione di human flourishing nella riflessione etica ed epistemologica contemporanea. In M. De Caro & C. Navarini (Eds.), *Forme e significati della fioritura morale* (pp. 11-58). Inschibboleth Edizioni.

Birkhäuer, J., Gaab, J., Kossowsky, J., Hasler, S., Krummenacher, P., Werner, C., & Gerger, H. (2017). Trust in the Health Care Professional and Health Outcome: A Meta-Analysis. *PloS one, 12*(2), e0170988. https://doi.org/10.1371/journal.pone.0170988

Bleidorn, W. (2009). Linking Personality States, Current Social Roles and Major Life Goals. *European Journal of Personality*, *23*(6), 509-530. https://doi.org/10.1002/per.731

Blum, L. (1994). *Moral Perception and Particularity*. Cambridge University Press.

Brady, M. S. (2018). *Suffering and Virtue*. Oxford University Press.

Branch W. T. Jr. (2015). A Piece of My Mind. The Ethics of Patient Care. *JAMA*, *313*(14), 1421-1422. https://doi.org/10.1001/jama.2015.1080

Branch, W. T. (2000). Supporting the Moral Development of Medical Students. *Journal of General Internal Medicine, 15*, 503-508. https://doi.org/10.1046/j.1525-1497.2000.06298.x

Buchanan, K., Newnham, E., Ireson, D., Davison, C., & Geraghty, S. (2022). Care Ethics Framework for Midwifery Practice: A Scoping Review. *Nursing Ethics*, *29*(5), 1107-1133. https:/doi.org//10.1177/09697330221073996

Bury, M. (1982). Chronic Illness as Biographical Disruption. *Sociology of Health & Illness, 4*(2), 167-182. https://doi.org/10.1111/1467-9566.ep11339939

Calhoun, L. G., & Tedeschi, R. G. (Eds.). (2006). *Handbook of Posttraumatic Growth: Research and Practice*. Psychology Press.

Campodonico, A. (2018). Why Wisdom Needs Fortitude (and Viceversa). *Teoria. Rivista Di Filosofia, 38*(2), 67-77. *https://doi.org/10.4454/teoria.v38i2.45*

Campodonico A., Croce, M., & Vaccarezza, M. S. (2017). *Etica delle virtù. Un'introduzione.* Carocci.

Campbell, A. V., & Swift, T. (2002). What Does it Mean to be a Virtuous Patient. *Scottish Journal of Healthcare Chaplaincy, 5*(1), 29-35. https://doi.org/10.1558/hscc.v5i1.29

Carel, H. (2007). Can I Be Ill and Happy?. *Philosophia, 35*(2), 95-110. https://doi.org/10.1007/s11406-007-9085-5

Carel, H. (2010). A Patient Journey: Lymphangioleiomyomatosis. *British Medical Journal (BMJ)*. http://dx.doi.org/10.1136/bmj.c848

Carel, H. (2011). Phenomenology and its Application in Medicine. *Theoretical Medicine and Bioethics, 32*(1), 33-46. https://doi.org/10.1007/s11017-010-9161-x

Carel, H. (2016a). *Phenomenology of Illness*. Oxford University Press.

Carel, H. (2016b). *II-Virtue Without Excellence, Excellence Without Health. Aristotelian Society Supplementary Volume*. Oxford University Press.

Carel, H. (2019). *Illness: The Cry of the Flesh*. Routledge.

Carel, H., & Kidd, I. J. (2020). Expanding Transformative Experience. *European Journal of Philosophy, 28*(1), 199-213. https://doi.org/10.1111/ejop.12480

Carron, P. (2014). A Case for Virtue: Aristotle's Psychology and Contemporary Accounts of Emotion Regulation. *Images of Europe. Past, Present, Future: ISSEI 2014 - Conference Proceedings*.

Cassell, E. (1991). *The Nature of Suffering and the Goals of Medicine*. Oxford University Press.

Centers for Disease Control and Prevention, U.S. Department of Health and Human Services. (2009). *Chronic Diseases: The Power to Prevent, the Call to Control*. Http://www.cdc.gov/nccdphp/publications/AAG/chronic.htm

Chapman, E. (2006). *Sacred Work: Planting Cultures of Radical Loving Care in America*. Greenleaf Book Group Llc.

Chochinov, H. M. (2001). Depression in Cancer Patients. *The Lancet Oncology*, *2*(8), 499-505. https://doi.org/10.1016/S1470-2045(01)00456-9

Cimatti, F. (2012). Quel dolore che non deve sapersi. Il linguaggio e il problema dell'esperienza estetica. *Rivista di estetica, 51*, 193-214. https://doi.org/10.4000/estetica.1410

Clark, D. (1999). 'Total Pain', Disciplinary Power and the Body in the Work of Cicely Saunders 1958-67. *Social Science & Medicine, 49*(6), 727-736. https://doi.org/10.1016/s0277-9536(99)00098-2

Clark, D. (2007). From Margins to Centre: a Review of the History of Palliative Care in Cancer. *The Lancet oncology, 8*(5), 430-438. https://doi.org/10.1016/S1470-2045(07)70138-9

Colombetti, E. (2009). To cure e to care in situazioni di fine vita: la proporzionalità e le cure palliative. *Persona y bioética, 13*(1), 9-19.

Cordova, M. J., Cunningham, L. L., Carlson, C. R., & Andrykowski, M. A. (2001). Posttraumatic Growth Following Breast Cancer: a Controlled Comparison Study. *Health Psychology, 20*(3), 176-185. https://doi.org/10.1037/0278-6133.20.3.176

Corsi, D. C., Turriziani, A., Cavanna, L., Morino, P., Ribecco, A. S., Ciaparone, M., Lanzetta, G., Pinto, C., & Zagonel, V. (2019). Consensus Document of the Italian Association of Medical Oncology and the Italian Society of Palliative Care on Early Palliative Care. *Tumori Journal, 105*(2), 103-112. https://doi.org/10.1188/16.ONF.747-753

Csikszentmihalyi, M. (1991). *Flow: The Psychology of Optimal Experience*. HarperCollins.

Curzer, H. J. (1993). Is Care a Virtue for Health Care Professionals?. *The Journal of Medicine and Philosophy: A Forum for Bioethics and Philosophy of Medicine, 18*(1), 51-69. https://doi.org/10.1093/jmp/18.1.51

D'Aquino, T. (1992-1993). *Le questioni disputate, voll. I, II, III, La verità*. Trans. R. Coggi & V. O. Benetollo. Edizioni Studio Domenicano.

Darnell, C., Gulliford, L., Kristjánsson, K., & Panos, P. (2019). *Phronesis* and the Knowledge-Action Gap in Moral Psychology and Moral Education: A New Synthesis? *Human Development*, *62*, 101-129. https://doi.org/10.1159/000496136

De Caro, M. & Vaccarezza, M. S. (2020). Morality and Interpretation: The Principle of Phronetic Charity. *Ethical Theory and Moral Practice*, *23*(2), 295-307.

De Caro, M., Vaccarezza, M. S., & Niccoli, A. (2018). *Phronesis* as Ethical Expertise: Naturalism of Second Nature and the Unity of Virtue. *Journal of Value Inquiry*, *52*(3), 287-305. https://doi.org/10.1007/s10790-018-9654-9

De Caro, M., Marraffa, M., & Vaccarezza, M. S. (2021). The Priority of *phronesis*: How to Rescue Virtue Theory from its Critics. In M. De Caro, M. S. Vaccarezza (Eds.), *Practical Wisdom* (pp. 29-51). Routledge.

De Caro, M., Navarini, C. & Vaccarezza, M. S. (2024). Why Practical Wisdom Cannot be Eliminated. *Topoi, 43*, 895-910. https://doi.org/10.1007/s11245-024-10030-1

De Caro, M., Bina, F., Bonicalzi, S., Brunetti, R., Croce, M., Kerusauskaite, S., Navarini, C., Ricci, E., & Vaccarezza, M. S. (2025). Virtue monism and medical practice. Practical wisdom as cross-situational ethical expertise. *The Journal of Medicine and Philosophy: A Forum for Bioethics an*

Philosophy of Medicine, 50(2), 80-92, https://doi.org/10.1093/jmp/jhae051
De Panfilis, L., Di Leo, S., Peruselli, C., Ghirotto, L., & Tanzi, S. (2019). "I Go into Crisis When…": Ethics of Care and Moral Dilemmas in Palliative Care. *BMC Palliative Care, 18*(1), 1-8. https://doi.org/10.1186/s12904-019-0453-2
De Vries, M., & Leget, C. J. (2012). Ethical Dilemmas in Elderly Cancer Patients: a Perspective from the Ethics of Care. *Clinics in Geriatric Medicine, 28*(1), 93-104. https://doi.org/10.1016/j.cger.2011.10.004
Dekkers, W., Uerz, I., & Wils, J. P. (2005). Living Well with End Stage Renal Disease: Patients' Narratives Interpreted from a Virtue Perspective. *Ethical Theory and Moral Practice, 8*(5), 485-506. https://doi.org/10.1007/s10677-005-8242-9
Descartes, R. (1999). *Discourse on Method and Meditations on First Philosophy*. Hackett Publishing.
Denzin, N. K. (1989). *Interpretive Biography*. Sage.
Dionigi, A., & Gremigni, P. (2010). *Psicologia dell'umorismo.* Carocci.
Diogenes Laertius. (1979). *Lives of Eminent Philosophers* (Vol. I-II). Trans. R. D. Hicks. Harvard University Press.
Doris, J. M. (2002). *Lack of Character: Personality and Moral Behavior*. Cambridge University Press.
Dreyfus, S. E. (2004). The Five-Stage Model of Adult Skill Acquisition. *Bulletin of Science, Technology & Society, 24*(3), 177-181. https://doi.org/10.1177/0270467604264992
Dreyfus, H., & Dreyfus, S. (1991). Towards a Phenomenology of Ethical Expertise. *Human Studies, 14*(4), 229-250.
Du Boulay, S. (1987). *A Movement for the Care of the Incurably Ill. Cicely Saunders: the Founder*. Jaka Books.
Dukes Holland, K., & Holahan, C. K. (2003). The Relation of Social Support and Coping to Positive Adaptation to Breast Cancer. Psychology & Health, *18*(1), 15-29 https://doi.org/10.1080/0887044031000080656
Ecclestone, C. & Crombez, G. (1999). Pain Demands Attention: A Cognitive-Affective Model of the Interruptive Function of Pain. *Psychological Bulletin, 125*(3), 356-366. https://doi.org/10.1037/0033-2909.125.3.356
Eleni, B., Evangelia, T., Stefanos, M., Elena, D., & Gouva, M. (2024). Humor and Quality of Life in Adults With Chronic Diseases: A Systematic Review. *Cureus, 16*(2). https://doi.org/10.7759/cureus.55201

Endler, N. S., & Parker, J. D. (1990). Multidimensional Assessment of Coping: A Critical Evaluation. *Journal of Personality and Social Psychology*, *58*(5), 844-854. https://doi.org/10.2307/2136617

Engster, D. (2005). Rethinking Care Theory: The Practice of Caring and the Obligation to Care. *Hypatia*, *20*(3), 50-74. https://doi.org/10.1111/j.1527-2001.2005.tb00486.x

Ericsson, K. A. (1996). The Acquisition of Expert performance: An Introduction to Some of the Issues. In K. A. Ericsson (Eds.), *The Road to Excellence: The Acquisition of Expert Performance in the Arts and Sciences, Sports, and Games* (pp. 1-50). Erlbaum.

Feltovich, P. J., Prietula, M. J., & Ericsson, K. A. (2018). Studies of Expertise from Psychological Perspectives. In K. A. Ericsson, R. R. Hoffman, A. Kozbelt, & A. M. Williams (Eds.), *The Cambridge Handbook of Expertise and Expert Performance* (pp. 59-83). Cambridge University Press. https://doi.org/10.1017/9781316480748.006

Ferrell, B. R., Temel, J. S., Temin, S., Alesi, E. R., Balboni, T. A., Basch, E. M., Firn, J. I., Paice, J. A., Peppercorn, J. M., Phillips, Y., Stovall, E. L., Zimmermann, C., & Smith, T. J. (2017). Integration of Palliative Care Into Standard Oncology Care: American Society of Clinical Oncology Clinical Practice Guideline Update. *Journal of Clinical Oncology*, *35*(1), 96-112.

Fleeson, W., & Gallagher, P. (2009). The Implications of Big Five Standing for the Distribution of Trait Manifestation in Behavior: Fifteen Experience-Sampling Studies and a Meta-Analysis. *Journal of Personality and Social Psychology*, *97*(6), 1097-1114. https://doi.org/10.1037/a0016786

Fleeson, W., & Jayawickreme, E. (2015). Whole Trait Theory. *Journal of Research in Personality*, *56*, 82-92. https://doi.org/10.1016/j.paid.2018.06.045

Fliedner, M., Zambrano, S. C., & Eychmueller, S. (2021). Public Perception of Palliative Care: a Survey of the General Population. *Palliative Care and Social Practice*, *15*, n. 263235242110175. https://doi.org/10.1177/26323524211017546

Folkman, S., & Lazarus, R. S. (1980). An Analysis of Coping in a Middle-Aged Community Sample. *Journal of Health and Social Behavior*, *21*(3), 219-239. https://doi.org/10.2307/2136617

Fowers, B. J., Richardson, F. C., & Slife, B. D. (2017). *Frailty, Suffering, and Vice: Flourishing in the Face of Human Limitations*. American Psychological Association.

Frankfurt, H. (2001). *The dear self.* Philosophers' Imprint.

Fredrickson, B. L., Tugade, M. M., Waugh, C. E., & Larkin, G. R. (2003). What Good are Positive Emotions in Crises? A Prospective Study of Resilience and Emotions Following the Terrorist Attacks on the United States on September 11th, 2001. *Journal of Personality and Social Psychology*, *84*(2), 365-376.

Fujimoto, T., & Okamura, H. (2021). The influence of coping types on post-traumatic growth in patients with primary breast cancer. *Japanese Journal of Clinical Oncology, 51*(1), 85-91. https://doi.org/10.1093/jjco/hyaa159

Gilligan, C. (1982). *In a Different Voice*. Harvard University Press.

Giordano, J., & Schatman, M. E. (2008a). An Ethical Analysis of Crisis in Chronic Pain Care: Facts, Issues and Problems in Pain Medicine; Part I. *Pain Physician, 11*(4), 483-490. https://doi.org/10.36076/ppj.2008/11/483

Giordano, J., & Schatman, M. E. (2008b). A Crisis in Chronic Pain Care: an Ethical Analysis. Part Two: Proposed Structure and Function of an Ethics of Pain Medicine. *Pain Physician, 11*(5), 589-595. https://doi.org/10.36076/ppj.2008/11/589

Giordano, J., & Schatman, M. E. (2008c). A Crisis in Chronic Pain Care: An Ethical Analysis. Part Three: Toward an Integrative, Multi-Disciplinary Pain Medicine Built Around the Needs of the Patient. *Pain Physician, 11*(6), 775-784.

Graber, D. R. (2009). Organizational and Individual Perspectives on Caring in Hospitals. *Journal of Health and Human Services Administration, 31*(4), 517-537.

Groopman, J. (2004). *The Anatomy of Hope: How People Prevail in the Face of Illness*. Random House.

Gross, J. J. (2002). Emotion Regulation: Affective, Cognitive, and Social Consequences. *Psychophysiology*, *39*(3), 281-291. https://doi.org/10.1017/s0048577201393198

Gulliford, L. (2020). Virtue in Positive Psychology. *Acta Philosophica: Rivista Internazionale di Filosofia*, *29*(1), 91-111. Https://doi.org/10.19272/202000701006

Gulliford, L., & Roberts, R. C. (2018). Exploring the "Unity" of the Virtues: The Case of an Allocentric Quintet. *Theory & Psychology, 28*(2), 208-226. https://doi.org/10.1177/0959354317751666

Guzek, Z., & Kowalska, J. (2020). Analysis of the Degree of Acceptance of Illness Among Patients After a Stroke: An Observational Study. *Clinical Interventions in Aging*, *15*, 2063-2072. https://doi.org/10.2147/CIA.S268095

Halwani, R. (2003). Care Ethics and Virtue Ethics. *Hypatia, 18*(3), 161-192. https://doi.org/10.1111/j.1527-2001.2003.tb00826.x

Harman, G. (2000). The Nonexistence of Character Traits. *Proceedings of the Aristotelian Society, 100*(1), 223-226. https://doi.org/10.1111/j.0066-7372.2003.00013.x

Heathwood, C. (2007). The Reduction of Sensory Pleasure to Desire. *Philosophical Studied, 133*(1), 23-44. https://doi.org/10.1007/s11098-006-9004-9

Heidegger, M. (1996 [1927]). *Being and Time* (Trans. J. Stambaug). State of New York University Press.

Held, V. (1993). *Feminist Morality: Transforming Culture, Society, and Politics*. University of Chicago Press.

Held, V. (2006). *The Ethics of Care*. Oxford University Press.

Heller, D. A., Watson, D. I., Komar, J., Min, J., & Perunovic, W. Q. E. (2007). Contextualized Personality: Traditional and New Assessment Procedures. *Journal of Personality*, *75*(6), 1229-1254. https://doi.org/10.1111/j.1467-6494.2007.00474.x

Hermsen, M.A., & Ten Have, H. A. M. J. (2003). Practical Ethics of Palliative Care. *American Journal of Hospice and Palliative Medicine*, *20*(2), 97-98. https://doi.org/10.1177/104990910302000206

Hillen, M. A., De Haes, H. C., & Smets, E. M. (2011). Cancer Patients' Trust in Their Physician-a Review. *Psycho□Oncology*, *20*(3), 227-241. https://doi.org/10.1002/pon.1745

Hume, D. (1958 [1739]). *A Treatise of Human Nature*. Clarendon Press (L. A. Selby-Bigge Eds.).

Hurka, T. (2000). *Virtue, Vice, and Value*. Oxford University Press.

International Association for Hospice and Palliative Care. (2018). *Global Consensus based Palliative Care Definition*. https://hospicecare.com/what-we-do/projects/consensus-based-definition-of-palliative-care/definition/

Jayawickreme, E., & Blackie, L. E. (2016). *Exploring the Psychological Benefits of Hardship: A Critical Reassessment of Posttraumatic Growth*. Springer.

Kahneman, D. (2011). *Thinking, Fast and Slow*. MacMillan.

Karami, A., Kahrazei, F., & Arab, A. (2018). The Role of Humor in Hope and Posttraumatic Growth Among Patients with Leukemia. *Journal of Fundamentals of Mental Health, 20*(3), 176-184.

Kassim, N. M., Sumari, M., & Razak, N. A. (2021). Surviving the Illness: A Phenomenological Study of Women Diagnosed With Cancer. *Malaysian Journal of Medicine and Health Sciences, 17*(4), 139-149.

Kauppinen, A. (2020). The World According to Suffering. In D. Bain, M. Brady, & J. Corns (Eds.), *Philosophy of Suffering Metaphysics, Value, and Normativity* (pp. 19-37). Routledge.

Kidd, I. J. (2012). Can Illness be Edifying?. *Inquiry, 55*(5), 496-520. https://doi.org/10.1080/0020174X.2012.716203

Kidd, I. J. (2015). Transformative Suffering and the Cultivation of Virtue. *Philosophy, Psychiatry, & Psychology, 22*(4), 291-294. https://doi.org/10.1353/ppp.2015.0048

Kirkengen, A. L., & Ulvestad, E. (2007). Heavy Burdens and Complex Disease – An Integrated Perspective. *Tidsskrift for Den Norske Legeforening.*

Kittay, E. F. (1999). *Love's Labor: Essays on Women, Equality, and Dependency*. Routledge.

Kittay, E. (2002). When Caring is Just and Justice is Caring: Justice and Mental Retardation. In E. F. Kittay & E. K. Feder (Eds.), *The Subject of Care* (pp. 257-276). Rowman & Littlefield Publishers.

Kittay, E. F., & Meyers, D. T. (1989). Women and Moral Theory. *Hypatia, 4*(2), 186-188.

Klein, C. (2015). What Pain Asymbolia Really Shows. *Mind, 124*(494), 493-516, https://doi.org/10.1093/mind/fzu185

Klitzman, R. (2008). *When Doctors become Patients*. Oxford University Press.

Kounti-Zafeiropoulou, F. (2014). The Contribution of Humor in Our Lives. *Studying Humour-International Journal,1*.

Kristjánsson, K. (2018). *Virtuous Emotions*. Oxford University Press.

Kübler-Ross, E. (2009 [1970]). *On Death and Dying. What the Dying Have to Teach Doctors, Nurses, Clergy and Their Own Families* (40th Ed.). Routledge.

Kübler-Ross, E., Wessler, S., & Avioli, L. V. (1972). On Death and Dying. *Journal of the American Medical Association, 221*(2), 174-179. https://doi.org/doi:10.1001/jama.1972.03200150040010

Lachman, V. D. (2012). Applying the Ethics of Care to Your Nursing Practice. *Medsurg Nursing, 21*(2), 112-116.

Lebacqz, K., (1985). The Virtuous Patient. In E. E. Shelp (Eds.), *Virtue and Medicine. Philosophy and Medicine* (Vol. 17, pp. 275-289). Springer. https://doi.org/10.1007/978-94-009-5229-4_14

Lebacqz, K. (1992). Humility in Health Care. *The Journal of Medicine and Philosophy, 17*(3), 291-307. https://doi.org/10.1093/jmp/17.3.291

Lee, Y. Y., & Lin, J. L. (2008). Linking Patients' Trust in Physicians to Health Outcomes. *British Journal of Hospital Medicine, 69*(1), 42-46. https://doi.org/10.12968/hmed.2008.69.1.28040

Lee, Y. Y., & Lin, J. L. (2011). How Much Does Trust Really Matter? A Study of the Longitudinal Effects of Trust and Decision-Making Preferences on Diabetic Patient Outcomes. *Patient Education and Counseling*, *85*(3), 406-412. https://doi.org/10.1016/j.pec.2010.12.005

Leiter, B. (2003). *Nietzsche on Morality*. Routledge.

Lindsey, E. (1996). Health within Illness: Experiences of Chronically Ill/Disabled People. *Journal of Advanced Bursing, 24*(3), 465-472. https://doi.org/10.1046/j.1365-2648.1996.02135.x

Lott, M. (2021). The Comic Sword of Truth and Justice: Humor as a Moral Virtue. In G. Pettigrove & C. Swanton (Eds.), *Neglected Virtues* (pp. 114-137). Routledge.

Lutfey, K. E., & Wishner, W. J. (1999). Beyond "Compliance" is "Adherence". Improving the Prospect of Diabetes Care. *Diabetes Care*, *22*(4), 635-639. https://doi.org/10.2337/diacare.22.4.635

MacIntyre, A. C. (1999). *Dependent Rational Animals. Why Human Beings Need the Virtues.* Duckworth.

Mandelbaum, M. (1969). *The Phenomenology of Moral Experience*. Johns Hopkins University Press.

Martínez-Martín, P., & Damián, J. (2010). Depression and Anxiety in Parkinson Disease. *Nature Reviews Neurology*, *6*(5), 243-245. https://doi.org/10.1038/nrneurol.2010.49

Martins Pereira, S., Fonseca, A. M., & Sofia Carvalho, A. (2011). Burnout in Palliative Care: a Systematic Review. *Nursing Ethics*, *18*(3), 317-326. https://doi.org/10.1016/j.mcna.2019.12.007

McAdams, D. P. (2011). Narrative Identity. In S. J. Schwartz, K. Luyckx, K., & V. L. Vignoles (Eds.), *Handbook of Identity Theory and Research* (pp. 99-115). Springer Science & Business Media.

McCullough, M. E., Emmons, R. A., & Tsang, J. A. (2002). The Grateful Disposition: a Conceptual and Empirical Topography. *Journal of Personality and Social Psychology*, *82*(1), 112-127. https://doi.org/10.1037/0022-3514.82.1.112

McLaren, M. (2001). Feminist Ethics: Care as a Virtue. In P. DesAutels, & J. Waugh (Eds.), *Feminists Doing Ethics* (pp. 101-118). Rowman & Littlefield Publishers.

Merleau-Ponty, M. (1962 [1945]). *Phenomenology of Perception.* Routledge.

Miles, J. K. (2019). Taking Patient Virtue Seriously. *Theoretical Medicine and Bioethics, 40*(2), 141-149. https://doi.org/10.1007/s11017-019-09484-x

Miller, C. B. (2021). *Honesty: The Philosophy and Psychology of a Neglected Virtue*. Oxford University Press.

Mittleman, A. (2009). *Hope in a Democratic Age: Philosophy, Religion, and Political Theory*. Oxford University Press.

Moch, S. D. (1989). Health within Illness. Conceptual Evolution and Practical Possibilities. *Advances in Nursing Science, 11*(4), 23-31. https://doi.org/10.1097/00012272-198907000-00006

Mortari, L. (2016). For a Pedagogy of Care. *Philosophy Study, 6*(8), 455-463. https://doi.org/10.17265/2159-5313/2016.08.001

Mortari, L., & Ubbiali, M. (2017). The "MelArete" Project: Educating Children to the Ethics of Virtue and of Care. *European Journal of Educational Research, 6*(3), 269-278. https://doi.org/10.12973/eu-jer.6.3.269

Mystakidou, K., Tsilika, E., Parpa, E., Galanos, A., & Vlahos, L. (2008). Post-Traumatic Growth in Advanced Cancer Patients Receiving Palliative Care. *British Journal of Health* Psychology, *13*(4), 633-646. https://doi.org/10.1348/135910707X246177

Narvaez, D. (2006). Integrative Ethical Education. In M. Killen, & J. Smetana (Eds.), *The Handbook of Moral Development* (pp. 703-733). Psychology Press.

Narvaez, D., & Lapsley, D. K. (2005). The Psychological Foundations of Everyday Morality and Moral Expertise. In D. K. Lapsley, & F. C. Power (Eds.), *Character Psychology and Character Education* (pp. 140-165). University of Notre Dame Press.

Navarini, C. (2018). La sofferenza nella malattia: lo sguardo antropologico. *Quaderni di scienza e vita, 18*, 33-37.

Navarini, C. (2020). *Cure palliative simultanee e sviluppo delle virtù*. Orthotes.

Navarini, C. (2022). La triade virtuosa di saggezza-speranza-fortezza nella malattia. In S. Langella, M. S. Vaccarezza, & M. Croce (Eds.), *Virtù Legge e Fioritura Umana. Saggi in onore di Angelo Campodonico* (pp. 177-204). Orthotes.

Navarini, C., & De Monte, E. (2019). Fear as Related to Courage: An Aristotelian-Thomistic Redefinition of Cognitive Emotions. *Humana. Mente, 12*(35), 167-189.

Navarini, C., & Ricci, E. (2021). La sofferenza di fine-vita: cure palliative simultanee, virtù e fioritura morale nella fragilità. *Bioetica, 4,* 401-413.

Niccoli, A., Piantoni, M., & Ricci, E. (2024). Virtue Monism. Some Advantages for Character Education. *Topoi*, 43, 1043-1051. https://doi.org/10.1007/s11245-024-10041-y

Nietzsche, F. (1994 [1887]). *On the Genealogy of Morality* (Trans. C. Diethe Eds.). K. Ansell-Pearson. Cambridge University Press.

Noddings, N. (1982). *Caring: A Feminine Approach to Ethics and Moral Education*. University of California Press.

Noddings, N. (2013 [1984]). *Caring: A Relational Approach to Ethics and Moral Education*. University of California Press.

Noddings, N. (2002). *Starting at Home. Caring and Social Policy*. University of California Press.

Noddings, N. (2005). *The Challenge to Care in Schools: An alternative approach to education*. Teachers College Press.

Panjwani, A. A., & Revenson, T. A. (2021). Coping with Chronic Illness. In K. Sweeny, M. L. Robbins, & L. M. Cohen (Eds.), *The Wiley Encyclopedia of Health Psychology* (pp. 61-70). John Wiley & Sons Ltd. https://doi.org/10.1002/9781119057840.ch51

Pankowski, D., Wydrychiewicz-Pankowska, K., Janowski, K., & Pisula, E. (2022). The Relationship between Primary Cognitive Appraisals, Illness Beliefs, and Adaptation to Rheumatoid Arthritis: A Systematic Review and Meta-Analysis. *Journal of Psychosomatic Research*, *164*, n. 11074. https://doi.org/10.1016/j.jpsychores.2022.111074

Park, C. L., Lechner, S. C., Antoni, M. H., & Stanton, A. L. (Eds.). (2009). Medical Illness and Positive Life Change: Can Crisis Lead to Personal Transformation?. *American Psychological Association*. https://doi.org/10.1037/11854-000

Park, C. L., Cohen, L. H., & Murch, R. L. (1996). Assessment and Prediction of Stress-Related Growth. *Journal of Personality, 64*(1), 71-105. https://doi.org/10.1111/j.1467-6494.1996.tb00815.x

Paul, L. A. (2014). *Transformative Experience*. Oxford University Press.

Pellegrino, E. D. (1988). *For the Patient's Good: The Restoration of Beneficence in Health Care*. Oxford University Press.

Pellegrino, E. D. (2007). Professing Medicine, Virtue Based Ethics and the Retrieval of Professionalism. In R. J. Walker, & P. J. Ivanhoe (Eds.), *Working Virtue: Virtue Ethics and Contemporary Moral Problems* (pp. 61-85). Oxford University Press.

Pellegrino, E. D., & Thomasma, D. C. (1993). *The Virtues in Medical Practice*. Oxford University Press.

Rachels, J. (1999). *The Elements of Moral Philosophy*. McGraw-Hill.

Ricci, E. (2020). Le virtù comunicative nella relazione medico-paziente. *Medicina e Morale, 69*(4), 503-521. https://doi.org/10.4081/mem.2020.855

Ricci, E. (2022). Buone pratiche di cura. Dalla Care Ethics ad una Virtue Care Ethics delle pratiche sanitarie. In S. Langella, M. S. Vaccarezza, & M. Croce (Eds.), *Virtù, legge e fioritura morale. Saggi in onore di Angelo Campodonico* (pp. 205-216). Mimesis.

Ricci, E. (2024a). Transforming Leaders to Transform Hospitals. Cultivating Ethical Leadership for Compassionate Healthcare. *Medicina e Morale, 73*(3), pp. 333-342. https://doi.org/10.4081/mem.2024.1595

Ricci, E. (2024b). Il ruolo della fioritura morale nella riappropriazione del sé disgregato. In M. De Caro & C. Navarini (Eds.), *Forme e significati della fioritura morale* (pp. 183-204). Inschibboleth Edizioni.

Ricci, E. (2025). Ristrutturare la percezione della malattia: l'importanza dello humor nelle malattie corniche. *Rivista Italiana di Studi sull'Umorismo, 8*(1), pp. 22-35.

Ricoeur, P. (1991). Narrative Identity. *Philosophy Today*, *35*(1), 73-81. https://doi.org/10.5840/philtoday199135136

Ruini, C., & Vescovelli, F. (2013). The Role of Gratitude in Breast Cancer: Its Relationships with Post-Traumatic Growth, Psychological Well-Being and Distress. *Journal of Happiness Studies*, *14*(1), 263-274. https://doi.org/10.1007/s10902-012-9330-x

Ruddick, S. (1982). Maternal Thinking. In A. C. Cafagna, R. T. Peterson, & C. A. Staudenbaur (Eds.), *Philosophy, Children, and the Family. Child Nurturance* (Vol. 1, pp. 101-126). Springer. https://doi.org/10.1007/978-1-4613-3473-6

Ruddick, S. (1989). *Maternal Thinking: Toward a Politics of Peace.* Ballantine Books.

Safran, D. G., Taira, D. A., Rogers, W. J., Kosinski, M., Ware, J. E., & Tarlov, A. R. (1998). Linking Primary Care Performance to Outcomes of Care. *Journal of Family Practice*, *47*(3), 213-220.

Sander-Staudt, M. (2006). The Unhappy Marriage of Care Ethics and Virtue Ethics. *Hypatia*, *21*(4), 21-40. https://doi.org/10.1111/j.1527-2001.2006.tb01126.x

Sander-Staudt, M. (2011). Care Ethics. Internet Encyclopedia of Philosophy. A Peer-Reviewed Academic Source. www.iep.utm.edu/care-eth/

Sanjo, M., Miyashita, M., Morita, T., Hirai, K., Kawa, M., Ashiya, T., Ishihara, T., Miyoshi, I., Matsubara, T., Nakaho, T., Nakashima, N., Onishi, H., Ozawa, T., Suenaga, K., Tajima, T., Hisanaga, T., & Uchitomi, Y. (2008). Perceptions of Specialized Inpatient Palliative

Care: A Population-Based Survey in Japan. *Journal of Pain and Symptom Management*, *35*(3), 275-282. https://doi.org/10.1016/j.jpainsymman.2007.04.021

Saunders, C. (1964) Care of Patients Suffering from Terminal Illness at St Joseph's Hospice. *Nursing Mirror, 14* February, vii-x.

Saunders, C. (1966). The Care of the Dying. *Guy's Hospital Gazette, 80*, 136-142.

Scoppettuolo, A. (2017). Vita morale ed etica della cura in Stan van Hooft. *Rivista triennale di scienza dell'amministrazione. Studi di teoria e ricerca sociale*, *9*(2), 1-34. Https://doi.org/10.32049/RTSA.2019.2.06

Schuchter, P., & Heller, A. (2018). The Care Dialog: the "Ethics of Care" Approach and its Importance for Clinical Ethics Consultation. *Medicine, Health Care and Philosophy*, *21*(1), 51-62. https://doi.org/10.1007/s11019-017-9784-z

Schnitker, S. A. (2012). An Examination of Patience and Well-Being. *The Journal of Positive Psychology*, *7*(4), 263-280. Https://doi.org/10.1080/17439760.2012.697185

Schnitker, S. A., Houltberg, B., Dyrness, W., & Redmond, N. (2017). The Virtue of Patience, Spirituality, and Suffering: Integrating Lessons from Positive Psychology, Psychology of Religion, and Christian Theology. *Psychology of Religion and Spirituality*, *9*(3), 264-275. https://doi.org/10.1037/rel0000099

Seoane, L., Tompkins, L. M., De Conciliis, A., & Boysen, P. G. (2016). Virtues Education in Medical School: the Foundation for Professional Formation. *Ochsner Journal, 16*(1), 50-55.

Shalev, A., Phongtankuel, V., Kozlov, E., Shen, M. J., Adelman, R. D., & Reid, M. M. (2018). Awareness and Misperceptions of Hospice and Palliative Care: A Population-Based Survey Study. *American Journal of Hospice and Palliative Medicine*, *35*(3), 431-439. https://doi.org/10.1177/1049909117715215

Slote, M. (1998a). The Justice of Caring. *Social Philosophy and Policy, 15*(1), 171-195. https://doi.org/10.1017/S0265052500003113

Slote, M. (1998b). Caring in the Balance. In J. Haber, & M. Halfon (Eds.), *Norms and Values* (pp. 27-35). Rowman and Littlefield Publishers.

Slote, M. (2007). *The Ethics of Care and Empathy*. Routledge.

Snow, N. E. (2010). *Virtue as Social Intelligence: An Empirically Grounded Theory*. Routledge.

Snow, N. E. (2013). Hope as an Intellectual Virtue. In M. W. Austin (Eds.), *Virtues in Action* (pp. 153-170). Palgrave McMillan. https://doi.org/10.1057/9781137280299_11

Snow, N. E. (2020). Is Hope a Moral Virtue?. In C. Blöser, & T. Stahl (Eds.), *The Moral Psychology of Hope* (pp. 171-188). Rowman & Littlefield.

Snow, N. E. (2018). Resilience and Hope. Invited panel presentation, *American Philosophical Association Pacific Division Meeting.*

Snow, N. E. (2025). *Hope and Resilience*. In N. L. King (Eds.), *The Virtues of Endurance* (pp. 82-107). Oxford University Press.

Snyder, C. R. (Eds.) (1999). *Coping: The Psychology of What Works.* Oxford University Press.

Snyder, C. R., Irving, L. M., & Anderson, J. R. (1991). Hope and Health. In C. R. Snyder, & D. R. Forsyth (Eds.), *Handbook of Social and Clinical Psychology: The Health Perspective* (pp. 285-305). Pergamon Press.

Sodergren, S. C., & Hyland, M. E. (2000). What Are the Positive Consequences of Illness?. *Psychology and Health, 15*(1), 85-97. https://doi.org/10.1080/08870440008400290

Solomon, R. (1999). Nietzsche's Virtues: A Personal Inquiry. *Royal Institute of Philosophy Supplement, 44*, 81-108. https://doi.org/10.1017/S1358246100006688

Stanton, A. L., & Revenson, T. A. (2011). Adjustment to Chronic Disease: Progress and Promise in Sesearch. In H. S. Friedman (Eds.), *Oxford Handbook of Health Psychology* (pp. 241-268). New York University Press.

Steyl, S. (2019). The Virtue of Care. *Hypatia, 34*(3), 507-526. https://doi.org/10.1111/hypa.12481

Stichter, M. (2018a). *The Skillfulness of Virtue, Improving our Moral and Epistemic Lives*. Cambridge University Press.

Stichter, M. (2018b). Virtue as a Skill. In N. E. Snow (Eds.), *The Oxford Handbook of Virtue* (pp. 57-84). Oxford University Press.

Stichter, M. (2020). Learning from Failure: Shame and Emotion Regulation in Virtue as Skill. *Ethical Theory and Moral Practice*, *23*(2), 341-354. https://doi.org/10.1007/s10677-020-10079-y

Stryker, S. (2007). Identity Theory and Personality Theory: Mutual Relevance. *Journal of Personality*, *75*(6), 1083-1102. https://doi.org/10.1111/j.1467-6494.2007.00468.x

Stuifbergen, A. K., Becker, H. A., Ingalsbe, K., & Sands, D. (1990). Perception of health among adults with disabilities. *Health Values*, 14(02), 18- 26.

Svenaeus, F. (2000a). Das Unheimliche-Towards a Phenomenology of Illness. *Medicine, Health Care and Philosophy, 3*(1), 3-16. https://doi.org/10.1023/A:1009943524301

Svenaeus, F. (2000b). *The Hermeneutics of Medicine and the Phenomenology of Gealth: Steps towards a Philosophy of Medical Practice* (Vol. 5). Springer Science & Business Media.

Swanton, C. (2015). *The Virtue Ethics of Hume and Nietzsche*. John Wiley & Sons.

Swift, T. L., Ashcroft, R. E., Tadd, W., Campbell, A. V., & Dieppe, P. A. (2002). Living Well Through Chronic Illness: the Relevance of Virtue Theory to Patients with Chronic Osteoarthritis. *Arthritis Care & Research, 47*(5), 474-478. https://doi.org/10.1002/art.10664

Tassinari, D., Drudi, F., Monterubbianesi, C., Stocchi, M., Ferioli, I., Marzaloni, A., Tamburini, E., & Sartori, S. (2016). Early Palliative Care in Advanced Oncologic and Non-Oncologic Chronic Diseases: a Systematic Review of Literature. *Reviews on Recent Clinical Trials, 11*(1), 63-71. https://doi.org/10.2174/1574887110666151014141650

Tedeschi, R. G., & Calhoun, L. G. (1996). The Posttraumatic Growth Inventory: Measuring the Positive Legacy of Trauma. *Journal of Traumatic Stress, 9*(3), 455-471. https://doi.org/10.1007/BF02103658

Tedeschi, R. G., & Calhoun, L. G. (2004). Posttraumatic Growth: Conceptual Foundations and Empirical Evidence. *Psychological Inquiry, 15*(1), 1-18. https://doi.org/10.1207/s15327965pli1501_01

Toombs, S. K. (2001). The Role of Empathy in Clinical Practice. *Journal of Consciousness Studies*, *8*(5-7), 247-258. https://doi.org/10.1093/iwc/iww010

Tracey Jr, W. D. (2017). Nociception. *Current Biology, 27*(4), R129-R133. https://doi.org/10.1016/j.cub.2017.01.037

Trachtenberg, F. L., Dugan, E., & Hall, M. A. (2005). How Patients' Trust Relates to their Involvement in Medical Care. *Journal of Family Practice*, *54*(4), 344-352.

Tran, S. K. (2017). GOOGLE: A Reflection of Culture, Leader, and Management. *International Journal of Corporate Social Responsibility*, *2*(10), 1-14. https://doi.org/10.1186/s40991-017-0021-0

Tronto, T. (1993). *Moral Boundaries: A Political Argument for an Ethic of Care*. Routledge.

Um, S. (2021). What is a Relational Virtue?. *Philosophical Studies*, *178*(1), 95-111. https://doi.org/10.1007/s11098-020-01422-1

Upton, C. L. (2009). Virtue Ethics and Moral Psychology: The Situationism Debate. *The Journal of Ethics*, *13*(2-3), 103-115. https://doi.org/10.1007/s10892-009-9054-2

Usami, H. (2012). The Effects of Proactive Coping Strategies on Psychological Well-Being. *Bull Seitoku Univ Jr Coll, 23*, 9-14.

van Hooft, S. (1995). *Caring: an Essay in the philosophy of Ethics*. University Press of Colorado.

van Hooft, S. (1998). Suffering and the Goals of Medicine. *Health Care and Philosophy, 1*(2), 125-131. https://doi.org/10.1023/A:1009923104175

van Hooft, S. (1999). Acting from the Virtue of Caring in Nursing. *Nursing Ethics, 6*(3), 189-201.

van Hooft, S. (2003). Caring and Ethics in Nursing. In V. Tschudin (Eds.), *Approaches to Ethics* (pp. 1-12). Butterworth Heinemann.

Van Nistelrooij, I., Visse, M., Spekkink, A., & De Lange, J. (2017). How Shared is Shared Decision-Making? A Care-Ethical View on the Role of Partner and Family. *Journal of Medical Ethics*, *43*(9), 637-644. http://dx.doi.org/10.1136/medethics-2016-103791

Vigani, D. (2017). Is Patience a Virtue?. *The Journal of Value Inquiry*, *51*(2), 327-340. https://doi.org/10.1007/s10790-016-9572-7

Wanzer, M., Booth-Butterfield, M., & Booth-Butterfield, S. (2005). "If we didn't use humor, we'd cry": Humorous coping communication in health care settings. *Journal of Health Communication*, *10*(2), 105-125. https://doi.org/10.1080/10810730590915092

Watson, J. (1997). The Theory of Human Caring: Retrospective and Prospective. *Nursing Science Quarterly*, *10*(1), 49-52. https://doi.org/10.1177/089431849701000l

World Health Organization. (2018), March 30. *Mental Health: Strengthening our Response*. https://www.who.int/news-room/fact-sheets/detail/mental-health-strengthening-our-response

Wright, J. C., Snow, N., & Warren, M. (2020a). Suffering, Virtue, and Character: Why the Science of Virtue Matters. *Acta Philosophica: International Journal of Philosophy, 29*(1), 55-72.

Wright, J. C., Warren, M. T., & Snow, N. E. (2021). *Understanding Virtue: Theory and Measurement*. Oxford University Press.

Zaner, R. M. (2005). A Work in Progress. *Theoretical Medicine, 26*(1), 89-104. https://doi.org/10.1007/s11017-004-4807-1

Zagzebski, L. (1996). *Virtues of the Mind*. Cambridge University Press.

Zagzebski, L. (2017). *Exemplarist Moral Theory*. Oxford University Press.

Zimmermann, C., Swami, N., Krzyzanowska, M. K., Leighl, N. B., Rydall, A., Rodin, G., Tannock, I. F., & Hannon, B. (2016). Perceptions of Palliative Care Among Patients with Advanced Cancer and their Caregivers. *Canadian Medical Association Journal*, *188*(10), E217-E227. https://doi.org/10.1503/cmaj.151171xf

MIMESIS GROUP
www.mimesis-group.com

MIMESIS INTERNATIONAL
www.mimesisinternational.com
info@mimesisinternational.com

MIMESIS EDIZIONI
www.mimesisedizioni.it
mimesis@mimesisedizioni.it

ÉDITIONS MIMÉSIS
www.cditionsmimesis.fr
info@editionsmimesis.fr

MIMESIS COMMUNICATION
www.mim-c.net

MIMESIS EU
www.mim-eu.com

www.ingramcontent.com/pod-product-compliance
Lightning Source LLC
Chambersburg PA
CBHW031346230926
51848CB00002B/408
9788869774973